To Laurie —
Very best wishes,
from John.

2019

LIGHTSPEED

Spacecraft Cassini-Huygens near Saturn, 2016.

LIGHTSPEED

The Ghostly Aether and the Race
to Measure the Speed of Light

John C. H. Spence

OXFORD
UNIVERSITY PRESS

OXFORD
UNIVERSITY PRESS

Great Clarendon Street, Oxford, OX2 6DP,
United Kingdom

Oxford University Press is a department of the University of Oxford.
It furthers the University's objective of excellence in research, scholarship,
and education by publishing worldwide. Oxford is a registered trade mark of
Oxford University Press in the UK and in certain other countries

First Edition published in 2020

Impression: 1

Published in the United States of America by Oxford University Press
198 Madison Avenue, New York, NY 10016, United States of America

British Library Cataloguing in Publication Data

Data available

Library of Congress Control Number: 2019937510

ISBN 978–0–19–884196–8

DOI: 10.1093/oso/9780198841968.001.0001

Printed and bound by
CPI Group (UK) Ltd, Croydon, CR0 4YY

For Archie, Colin, and Andrew.

Contents

Introduction

This book describes a great scientific adventure and the extraordinary personalities who were involved. It is the adventure, founded in human curiosity, and starting from the time of the ancient Greeks, which has given us the mobile phone, the internet, television, satellite communication, GPS, Einstein's theory of relativity, and our understanding of the cosmos and man's place in it. It is the story of the measurement of the speed of light.

Three enormously difficult and important ideas lie at the heart of this story, ideas which mankind has grappled with for centuries. Firstly, that light does not travel instantaneously from the stars, so that we are looking back in time when we see them. Secondly, that light (and radio waves) can travel in a complete vacuum, with nothing at all present to support their propagation at constant speed. And finally, if we take that constant speed seriously, as Einstein did, we are led inexorably to his famous equation for the amount of energy released by an atom bomb. This book will take you through the development of those ideas in the simplest way possible, in roughly chronological order.

The book is thus the story of efforts by philosophers and scientists throughout the ages to answer the questions: What is light? What is electricity? How can light, radio waves, and even gravity waves travel through complete vacuum? Do gravity and light act instantaneously or take time to travel? And if they do take time, would people on remote planets orbiting distant stars see dinosaurs if they looked at Earth? Why is the speed of light, which takes more than a tenth of a second to circle the Earth (and an hour to get here from Saturn), the highest speed at which any *object* can travel? How can control signals be sent to space probes like *Cassini* at Saturn, with these hour-long delays? Does quantum mechanics allow us to send signals faster than lightspeed, and what role does Einstein's relativity play in all this? I have tried to give simple, clear answers to all these questions, alongside the very human story of the struggles, both conceptual and experimental, of the great

Lightspeed: The ghostly Aether and the race to measure the speed of light. John C. H. Spence.
© John C. H. Spence 2020. Published in 2020 by Oxford University Press.
DOI: 10.1093/oso/9780198841968.001.0001

physicists who have given us our current understanding, based on my reading and experience teaching related courses in physics.

The story will take us from the ancient Greek astronomers, and all their efforts to measure the solar system and the Earth, to Fermat and Descartes, then to the remarkable Ole Roemer, the Danish astronomer who noticed, using Galileo's newly invented telescope, that the eclipses of Jupiter's moon Io were sometimes a bit late. From this it was possible to estimate the speed of light reasonably accurately for the first time in around 1664. Captain Cook's adventures in Tahiti at the time of the transit of Venus are important to our story, as is James Bradley, who did perhaps the simplest and best experiment of all, one of great importance to Einstein in his paper announcing the theory of relativity in 1905. And so to the great physicists of the nineteenth century, Kelvin, and the genius, almost a magician, James Clerk Maxwell, and his followers, FitzGerald, the brilliant and generous Lorentz, the reclusive and eccentric Oliver Heaviside. Then the supremely gifted experimentalists Heinrich Hertz, who discovered radio, Albert Michelson in his search for the Aether wind, and Fizeau and Foucault working in Paris to measure the speed of light in the laboratory. And finally Einstein and the ideas leading up to the use of quantum effects to communicate faster than the speed of light.

How fascinating, to have been at the dinner table in Baltimore in 1884, where the young Michelson in despair told the visiting Lords Rayleigh and Kelvin that, because of his "failed" experiment in Berlin and the lack of any response to his paper, he was giving up trying to detect the Aether wind which they all firmly believed in, that God-given absolute frame of reference at rest in the universe, which supported light waves, and through which our Earth was zooming along at 67,000 miles per hour around the Sun. Rayleigh and Kelvin convinced him to try just once more in his new position at Case-Western University, convinced that he would detect the wind with a more sensitive instrument. That he did, undertaking one of the most important experiments in the history of physics (which again failed to detect an Aether), giving a result that no one wanted, and only Einstein could fully understand. In so doing, he created a new theory in which the speed of light played a crucial role, which changed our understanding of the very nature of time itself. As a result, we'll learn how light travels as a wave, but arrives as a particle.

Along the way we will meet fascinating characters who led extraordinary lives. Francois Arago, who escaped capture and imprisonment by pirates in north Africa while attempting to map out longitude for

Napoleon, and who got his friend Augustin Fresnel through his final exam (which he was about to fail) by demonstrating that there was indeed a bright spot in the center of a shadow, and then went on as a politician to eliminate slavery in the French colonies. And his friend Foucault, who also built a huge pendulum in a cathedral which shows directly the Earth's rotation. And Marie Cornu, shooting beams of light across the rooftops of Paris, bounced back by mirrors, soon after the bombardment and starvation of the population into surrender by Bismarck in 1871. (Figure 5.9 shows a menu in a posh Paris restaurant at the time, offering "Fricasee of Rats and Mice" to the famished inhabitants). The American Ambassador to France used the menu, braving the siege to act as negotiator.

The speed of light is one of a very small number of fundamental constants in physics which truly determines the nature of our universe and the form of matter within it. It is the constant c in Einstein's most famous equation, $E = mc^2$, relating energy release E (for example in nuclear weapons) to mass m, and has always been intimately connected with problems in navigation, from longitude determination to the modern global positioning system (GPS). Its measurement drove advances in technology, notably in interferometry and astronomical instruments. The speed of light has been described by S.R. Filonovich as the constant which provides "a clear manifestation of the unity of our physical world." And the discovery that light does not travel instantaneously tells us, as we look up into the night sky at distant stars, that we indeed are looking back in time. (Later, the stars move. Today they are somewhere else!) The history of the measurement of the speed of light follows one of the greatest intellectual adventures in human history, at the heart of progress in science over the past 400 years since Newton, and central to the later wave–particle duality, the idea that light can be thought of as either a wave, or a stream of tiny bullets.

This is also the story of some of the greatest scientific instrument makers the world has seen, leading up to today's technology of atomic clocks and GPS for navigation and control of driverless cars.

I've included some simple mathematics, which are not essential to understanding our story. The appendices contain more mathematics, and also describe a simple way for you to measure the speed of light using a microwave oven and some pizza dough. All the books and articles referred to in this book are listed at the end in the "Sources and References" section.

1

Early Ideas

In July 1671, a twenty-seven-year-old student left his studies at Copenhagen University for the short journey to Uraniborg ("The city of heaven") on the island of Hven, in the narrow waters between Sweden and Denmark. With Ole Roemer were a famous mathematics professor and a visiting astronomer from Paris. With their recently invented telescopes, they were visiting the abandoned observatory established long before by the great astronomer Tycho Brahe, the site of a present-day museum. The purpose of the trip was to obtain more accurate coordinates for the location of Brahe's laboratory. As a result, using both Brahe's and new observations, they hoped to test an idea of Galileo's—that watching the orbits of the moons of Jupiter by telescope could provide a universal clock for our planet. Since this could be seen at most places and times on Earth, it could therefore solve the long-standing problem of longitude determination. Poor navigation, due to lack of knowledge of longitude, had cost many lives at sea, and its determination had obvious military value. Philip of Spain had established a prize for a solution to this problem in 1598. This trip would completely change the life of Roemer and the course of scientific history, leading to a foundational principle for Einstein's theory of relativity, and measurement of the most important number in all of physics—the speed of light. But before we describe Roemer's adventures, which firmly established the modern view that light has a finite speed, we must summarize the events leading up to this turning point, to get it into historical perspective.

This history of increasingly accurate measurements of this fundamental physical constant by ingenious methods (such as rapidly rotating mirrors and atomic clocks) will lead us deep into questions which physicists have struggled with for centuries, such as *how can star-light travel through a complete vacuum? Is there really something there?* And, if there is some invisible, ghostly *Aether*, an invisible "vortex foam" which was invented

Lightspeed: The ghostly Aether and the race to measure the speed of light. John C. H. Spence.
© John C. H. Spence 2020. Published in 2020 by Oxford University Press.
DOI: 10.1093/oso/9780198841968.001.0001

to support light waves as water supports ocean waves, does it rotate and move with the Earth? Or is it stationary, and if so, with respect to what—to the most distant stars? Does it provide a God-given absolute frame of reference, the deepest question Einstein struggled with—a question which was only finally resolved through measurements of the speed of light? And how can the distance between the Earth and the Sun be measured; which is surely needed to obtain the velocity of light from the Sun?

The idea that outer space must be filled with some invisible stuff for light waves and gravitational forces to travel through goes back a long way. In his introduction to Heinrich Hertz's book *Electric Waves*, on his discovery of radio waves, the great British physicist Lord Kelvin wrote (in 1893):

> To fully appreciate this work now offered to the English reading public, we must carry our minds back two hundred years to the time when Newton made known to the world the law of universal gravitation. The idea that the sun pulls Jupiter, and Jupiter pulls back against the sun with equal force, and the sun, earth, moon and planets all act on one another with mutual attractions seemed to violate the supposed philosophical principle that matter cannot act where it is not

We can see from this how reasonable it was for physicists, right up into the early twentieth century, to believe that space must be filled with *something*—they called it the Aether—which could transport the light arriving on Earth from the stars. Until about the time of Roemer, this invisible medium was generally believed to transport light instantaneously. Newton became convinced by Roemer's results that light took time to travel, but believed that gravitational forces were nevertheless able to act instantaneously across vast distances, a theory which came to be known as *action at a distance*.

As a teenage schoolboy in the early 1960s, I was taught geometry from a textbook based on the Greek mathematician Euclid's axioms in his *Elements* (360 BCE), using their original numbering, and translated from the original Greek. That a person living so long ago could collect together ideas which teenagers would learn long into the future is truly remarkable—a kind of time travel, in which a message is sent thousands of years into the future. (A similar thing occurred when the melody *Greensleeves* was composed—the medieval composer could not have suspected that at any moment now a small fraction of the human

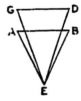

Figure 1.1 Euclid's diagram showing that closer objects appear larger. (From Burton (1945.))

population has this melody running through their heads. In this case—as at the opening of the Star Wars movie—the melody serves to establish a mood, which the composer manages to convey to listeners today across the centuries.)

Euclid's book is a summary of what was known at the time, and gives simple relationships between the dimensions of regular two- and three-dimensional shapes. These include Pythagoras's great theorem, giving the distance along the diagonal of a rectangular field, from measurements of the lengths of its two sides. He derives his results systematically from a series of axioms, setting the pattern for logical argument. We now know that this Euclidean geometry is only one of many possible geometries; later, the alternatives became important in connection with Einstein's work.

For anyone to write a book, that has been used more or less unchanged for more than two thousand years, is an extraordinary achievement. But Euclid of Alexandria wrote a second, much less well known book, *Optics*, which describes his theory of light using the same deductive method as his more famous *Elements*. This uses the results in his *Elements* to analyze perspective, and the apparent reduction in the size of objects as they move further away. His theory is based on the erroneous idea that human vision is due to rays (straight lines) which originate in the human eye (a kind of "visual fire"), projected toward the objects we look at. (In fact, the eye receives rays originating from the sun, or a lamp, which have reflected off the object we look at and into our eyes in straight lines). Euclid must have had a hard time explaining why we cannot see in the dark!

Referring to Figure 1.1, here is what Euclid, writing about 2300 years ago, says about perspective, and the way in which closer objects appear larger (translated from the Greek by H. Burton in 1945):

Objects of equal size unequally distant appear unequal and the one lying nearer to the eye always appears larger. (Fig. 5.)
Let there be two objects of equal size, *AB* and *GD*, and let the eye be indicated by *E*, from which let the objects be unequally distant, and let *AB* be nearer. I say that *AB* will appear larger. Let the rays fall, *EA*, *EB*, *EG*, and *ED*. Now, since things seen within greater angles appear larger, and the angle *AEB* is greater than the angle *GED*, *AB* will appear to be larger than *GD*.

By understanding the importance of the *angular spread* of an object (such as a dinosaur seen in the rear-view mirror) in determining its apparent size, Euclid had discovered an important principle, to which we will return in many different contexts. But he did not discuss how fast the light travelled along these rays, and it was generally assumed that light propagation was instantaneous.

The Greek philosopher Empedocles of Acragas (490–35 BC) may have been the first to suggest that light has a finite velocity, since he wrote of it "travelling" from the stars to Earth. The philosopher Hero of Alexander (in the first century) argued that, since the stars are immediately visible upon opening our eyes, and since light must travel from our eyes to stars and back, it must propagate instantaneously. In the second century, Ptolemy produced a book on optics (now lost, known from later quotations, and the basis for the later Arab scientific renaissance in this field in the tenth century), which extended Euclid's ideas, and which contained the correct ray diagrams for reflection from a mirror (axial rays only) and refraction. The Islamic scholars Abdallah Ibn Sina (980–1073) and Al-Hasan Ibn Al-Haitham, great physicists and specialists in optics at the height of medieval philosophy, both believed that light travelled with finite speed. Roger Bacon (1214–92) knew of Haitham's work and considered it important for astrology, since astrologers naturally needed to know whether the influence the stars had on our fortunes and behavior in life was transmitted instantaneously from them, or was a little delayed (!).

With the development of better lenses for telescopes, it became important to understand how they worked. Yet none of this research in optics had provided the correct ray diagram for imaging using a lens, such as the human eye, in order to explain image formation. Here, as shown in **Figure 1.4** (if we ignore the mirror and S), all rays diverging from a point S″ are gathered together by the lens L to recombine at the point S′. Then, for the eye (or a camera or telescope), S′ would be the image point on the retina (or detector) of one point S″ on an object being viewed at S″. The first scientist to understand this appears to have been Kepler, who shows similar diagrams in his books on optics, *Paralipomena* in 1604 (showing the correct ray diagram for a spherical glass lens) and *Dioptrice* in 1611, in explanation of the functioning of the lenses in Galileo's telescope. While he did not derive the "lens laws" taught to undergraduate science students, he does indicate an understanding of the formation of virtual images, which arise with mirrors, as we will discuss further. These books contain the earliest correct ray diagrams from which the subject of geometric optics was later developed. (You can read much more on the fascinating history of optics in Olivier Darrigol's (1912) excellent book on the subject, listed, like all the books and articles mentioned, in References at the end of this book.) Kepler also argued that since space is empty, it can offer no resistance to the propagation of light, which must therefore travel infinitely rapidly. Serious doubt was cast on this idea of instantaneous "action at a distance" in the seventeenth century. A detailed discussion of the issue first appeared in a book by the famous Italian scientist Galileo Galilei (1564–1642) in 1638. Here Galileo speculates that, if light does have a finite velocity, then its speed could be determined by uncovering a lantern on a hilltop at night, while viewing this from another hill with his newly invented telescope. The viewer would then immediately uncover another lamp, viewed now by the first sender. (Clearly, a mirror on the second hill would have been a much better arrangement, as later suggested by Descartes!). This experiment was actually undertaken by the Florentine Academy in 1667 (one of the first scientific societies) and published. They used a distance of about a mile between hills, but the result was inconclusive. Since we now know that the speed of light is about 186,000 miles per second, the failure of his experiment is not surprising! (A table of accurate values for the speed of light, the size of the solar system and the Earth, and other useful numbers, is given in Appendix 1). Galileo went on to discuss the observation of lightning flashes between clouds, which may be miles apart. Again he was unable to detect a delay between the start and the end of a flash.

Many factors contributed to the great scientific revolution which started in the sixteenth century, and the attitudinal changes which supported a rational basis for physical phenomena. These certainly included the invention of the printing press, which greatly assisted scholarly communication. The Protestant Reformation in the early sixteenth century also promoted attitudes of empiricism and a belief that nature could be understood using simple conceptual models.

A leader in this change was the great French philosopher Rene Descartes (1596–1650), who provided us with the next clue toward understanding the speed of light, and the great French mathematician and lawyer Pierre Fermat (1607–65), shown in **Figure 1.3**. Both attempted to understand the phenomenon of refraction. This occurs when a beam of light appears to bend as it enters water. Try this, by shining a laser pointer into a bowl of water. The problem of predicting the angle of the bend turns out to be equivalent to that of a lifeguard at the beach who sees a person drowning. Like light in a vacuum, they can run much more quickly on the beach than they can swim in water, and likewise, light slows down when it enters the water. In Figure 1.2, the lifeguard could be at A, entering the water at O, with the drowning person at B. At what point O should they enter the water to get to the swimmer at B in the shortest time? The solution to that problem applies also to the refraction problem.

We will see that this simple effect and its analysis, studied since the time of the Greeks, contains a wealth of fundamental physics, from the wave–particle duality to the path-integral methods of field theory and the action principle of mechanics. The study of this simple effect has been one of the most fruitful in the history of physics. Through their work, Descartes and Fermat will introduce us to the fierce debate among scientists from that time onwards as to whether light is a series of particles—tiny bullets—or a wave. We will see that in the modern view it is both. Light travels as a wave, but arrives as a particle.

Refraction is also responsible for the fact that a straight stick, thrust into clear water at an angle inclined to the surface, appears bent at the point where it enters the water. A formula had been known empirically for several hundred years, which accounts for this effect, better known today as *Snell's Law*, after the Dutch astronomer Snellius, who rediscovered it in 1621. (The correct form was first given by the Persian scientist Ibn Sahl in around 984.) It will be simplest if I describe our modern explanation first, since this will help us to understand the erroneous but important approach of Descartes and his followers—skip this if you have not done any introductory optics. We define the ratio of the speed of light in vacuum c to that v_2 in a medium, such as water or glass, as the refractive index n. (These speeds are actually *phase velocities*, which we will discuss later.) For water, n = 1.33, meaning that light travels 33% faster in vacuum than in water. The angle through which a light ray changes direction when it enters a new medium, as shown in Figure 1.2, is given by Snell's law,

$$\sin \theta / \sin \phi = c / v_2 = n.$$

We obtain this result by the following (symmetry) argument. Light can be described by a wavevector **K**, which points in the direction of light propagation, whose length is given by the reciprocal of its wavelength. Along the direction x (from O to C) there is no change in structure—all the variation in refractive index occurs along the z direction from O to E. So there can be no scattering or change in the magnitude of the x component K_x of the wavevector of the light. This means that OC = OD, and hence the wavevector of the light in the water must terminate on the line CB, which preserves this condition. Since the light is observed to bend toward the normal OE when entering the denser medium (water), this is only possible if the wavevector K′ of the light in the water becomes longer, that is, if the wavelength λ (the reciprocal of K′) of the light becomes shorter. Now it is well known from the study of sound and water waves that the frequency f, wavelength λ, and speed v (phase velocity) of a wave are related by $v = f\lambda$. This would apply to a vibrating guitar string....velocity v. The length of the string is the

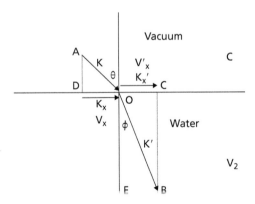

Figure 1.2 Snell's law. A ray of light at A in vacuum passes into water or glass with refractive index n toward B. The speed of the light in the glass is v_2.

wavelength, and the frequency (in cycles per second) is given by a guitar tuner device. The frequency of the light, which fixes its color, is seen not to change on entering water, so the shorter wavelength in the water must mean reduced speed, to keep the frequency constant. *The wavelength and speed adjust their values to keep the frequency (color) unchanged.*

In Newton's famous experiment in which he split sunlight into its constituent colors using a prism, he was using the fact that the refractive index of the prism changes with the frequency of light, so that rays at different frequencies were bent through different angles, and the colors separated.

Descartes' 1637 derivation reached the opposite conclusion, using an argument which, in his original manuscript is difficult to follow; however his conclusion is most important since it influenced generations of scientists and was consistent with Newton's views, which held great authority, supporting the particle view of light. He considers a particle of light crossing into, say, flat water at some angle to the vertical, and makes the analogy with a tennis ball puncturing and crossing a horizontal sheet of very thin fabric at an angle. He argues that this inclined collision should affect the vertical component of the *velocity* of the ball (instead of the wavevector) but not its horizontal component. (This might also be suggested by watching a stone thrown into water at a low angle, which, if it does not skip, sinks more vertically after hitting the water. But it slows down in the water.) Repeating the above argument from **Figure 1.2**, but replacing wavevectors by velocity vectors, we then find (incorrectly) that the velocity of light is greater in the denser water. If we stick to the definition of refractive index as the speed in vacuum (or air) divided by the speed in water, we then have the incorrect result that the refractive index for water is smaller than unity, but by inverting this definition we might have a usable result.

For the subsequent two centuries, scientists were then divided into those who, following Newton and Descartes, believed that light was a particle which travelled faster in denser media, and the growing view (most firmly established by Huygens, Fresnel, and Young) that it was a wave, which slowed down in water or glass. The most important of these papers by Huygens, Young, and Fresnel can be found in the collection edited by H. Crewe.

The importance of Descartes' idea is that it suggests that light does not travel instantaneously and must have a finite velocity, *in order that the velocity can change on crossing a boundary.* When using the correct value of n for the ratio of the velocity of light in vacuum to that in water (where it actually slows down), his treatment, which is not given using trigonometric functions, does predict the correct bending angle for the light. Descartes in later life became convinced that the velocity of light was actually infinite, but was able to use his theory of refraction in a brilliant application to explain the appearance of the rainbow. (The refractive index n can be treated as a useful constant, without worrying about its interpretation in terms of light speed.) His work supported the particle view of light, resurrected a usable form of Snell's law, and gave support to the idea that light travels faster in denser media.

Pierre Fermat (**Figure 1.3**) was born in Beaumont-de-Lomagne (the house is now a museum) and trained as a lawyer at the University of Orleans. He became a counselor at the Parlement de Toulouse, a position he held until his death. He spoke six languages and wrote copious poetry, but he is best known for his "last theorem" (only recently proven) and his contributions to number theory and analytic geometry.

Figure 1.3   Pierre Fermat (1607–65). (From Google Images.)

For Fermat, his real job was the law—mathematics was a hobby. He gave the results of most of his work (often without proof) only to friends in letters.

Fermat was able to derive the law for the sharp bending of light rays when they enter a denser medium in 1662, by assuming that light travels by the path of least resistance. This again suggested that it must have a finite velocity (in order to be affected by resistance), and so slow down in the denser medium (water). Fermat equated the path of least resistance with the path which takes least time, and so was able to derive the correct relationship between the speed of light in the two media (air and water, or air and glass, for example) and the angular deflection, which could be measured experimentally. This is the same problem as our lifeguard running and swimming (how much of each?) to save a drowning person.

Fermat may have known that Hero of Alexander had pointed out that, for light reflected from a mirror, the path (known as *specular*) is the shortest path, and adapted this idea to the shortest time for refraction. The proof of Fermat's least-time result requires some calculus, but we can understand the simpler and very similar case of a mirror from **Figure 1.4**. Fermat's principle also explains how mirrors work.

Here, light rays spread out a little (dotted lines, not dashed lines) from a very small lamp at S, and bounce off the mirror (making equal angles θ) at A, to enter the lens L of an eye, where they are focused onto the point S′ on the retina. Notice how the lens gathers the diverging rays from the lamp (after reflection from the mirror) and makes them

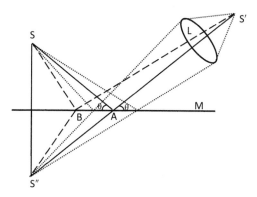

Figure 1.4 For a mirror, light rays travel by the shortest path, which fixes the point A where they bounce off a mirror M.

converge to a point on the retina. The mirror works because these dotted rays appear to come from S″. The person is looking in that direction, so the lamp S appears to be behind the mirror—the brain cannot tell the difference between those rays reflected from the mirror and those which would reach the eye if you looked directly at the lamp. If you looked across, you would see the real lamp at S, but instead you see its image reflected in the mirror at S″, as if there were no mirror at all. But there is no actual light or lamp behind the mirror.

We need to compare this true path from S via A to S′ with an alternative, incorrect path which bounces off the mirror at, say, B, shown dashed (not dotted), to find out which path takes the least time. The correct path SAS′ is said to undergo *specular* reflection, meaning that the incoming and reflected wave directions make equal angles with the surface of the mirror. Fermat claims the fastest trip goes via A, not B. In this simple case, the duration of the trip is proportional to the distance travelled, and all paths are taken at the same speed, so the fastest trip (unlike the bodyguard problem, with its two different speeds) will also be the shortest. Now, for any position of B along the mirror, the length of the real path SBS′ is equal to the length of the "virtual" path S″BS′, and we want to find the location of B which makes this path the shortest. Since the shortest path for S″AS′ is obviously through A (where it is a straight line), this must also be the shortest (and quickest) path for SBS′ (move B to A), and this is the path which light rays actually follow when reflected from a mirror, making equal angles θ (specular condition) at incidence and reflection. So Fermat has explained the specular (equal angle) condition which makes the mirror work. Note in particular that the spread-out rays (shown dotted) are also reflected with equal angles, so that the lens can form an image.

Christiaan Huygens included the correct geometric construction for Snell's law in his book *Trait de la Lumiere* in 1678. Johann Bernoulli (1667–1748), adopting Newton's proposal that light consisted of a flow of particles, later pointed out in 1701 that if the rays of light refracted at an interface are treated as the different force vectors acting on a particle at the interface, then Snell's law gives the required condition for force equilibrium (balance). His son Johann Bernouli (1710–90) extended these ideas in 1736 in one of the first theories of the luminiferous Aether, the invisible elastic medium which was supposed to fill the vacuum of outer space, in order to support transmission of light through space from the stars. The properties required of this ethereal medium were remarkable. The velocity of a wave in a medium is usually taken to be equal to the square root of the elastic modulus (*Young's modulus*) divided by the density. To support waves travelling at the enormous speed of light, the Aether, if it had an elasticity similar to steel, would require a density about 50,000 times less than that of hydrogen gas. Conversely, it might have a density similar to steel with an elastic modulus 3600 million times that of steel. And it needed to be invisible and to permeate all forms of matter and offer no resistance to the motion of planets. The great nineteenth-century physicist Lord Kelvin described it as "the only form of matter about which we know nothing at all." I have a copy of Kelvin's famous "Baltimore Lectures," given in 1884 at Johns Hopkins University, in which he estimates the density of the Aether to be about 10^{-20} pounds per cubic foot! This invisible "vortex sponge" was to be used by James Clerk Maxwell a century later to derive his famous equations describing the transmission of light through space.

By the late seventeenth century some respected natural philosophers or scientists (a word which did not then exist) accepted that refraction could best be explained by assuming that light travelled at a finite speed, and was not instantaneous.

It is useful to pause here to look back on how the correct theory of refraction was finally obtained, as an example of how theoretical physicists work. Prior to the time of Euler (who introduced the sine and cosine terminology in around 1748), physical theories, such as those of Newton, were given in terms of geometric constructions and simple equations derived from them. For refraction, scientists already knew the answer before they started, since Euclid's geometric constructions had been used previously to fit the observed ray paths over a range of incident angles, giving Snell's law. However, there was no physical model to account for the constant we now know as the refractive index. There was no mechanism or underlying physical basis for the effect. Descartes, perhaps from observation of heavy stones thrown into water at an angle, was willing to consider a finite velocity for light, a huge conceptual leap. Fermat accepted this also, and was led from the idea of greater *resistance* to the passage of light in a denser medium to the idea that it will travel by the path of least time, an inspired guess with huge later implications for physics.

A combination of guesswork (physical intuition) and willingness to be guided by experimental observations is the way physics normally progresses, however there are exceptions. The very great physicist Paul Dirac believed that the intrinsic beauty and symmetry of equations was of primary importance in guiding discoveries, experimental results being secondary to this. Einstein is often quoted as saying that "imagination is more important than knowledge," knowledge being a mere collection of facts. Historians of science have commented on the fact that Einstein's least productive period, working on unified field theory, involved his most advanced mathematical techniques, and his greatest earlier work, the least. Sabine Hossenfelder, in her book *Lost in Math: How beauty leads physics astray* suggests that the search for beauty in mathematics is currently misleading sub-atomic particle physics, and gives a good review of how the criteria of simplicity, naturalness, and elegance are used to select theories. In this field, new data are scarce (limited by the highest energy possible at the CERN particle accelerator) and competing theories (which must also fit a mountain of old data) are plentiful.

My own feeling as an experimentalist is that you can never know too much of the current state of theory, since this informs and guides one's physical intuition based on experimental evidence. And the theory can suggest which experiments are most likely to be productive. This approach may change as we enter an era of big data and ever more powerful computer simulation and modelling, using machine learning to predict outcomes or even formulate laws. However, it is significant that recent studies show that having more data to help make predictions is less important than having a good model (the *theory* or guiding principle on which an algorithm is based).

Fermat's method has since taken on much greater importance in physics, where it is known as the variational method, an idea which in simpler form again goes back to Euclid. It is also known as the principle of least (or stationary) action. It was fundamental to Richard Feynman's new formulation of quantum mechanics, the path integral method, in the 1950s, following earlier work by Paul Dirac in 1933. In this approach, we can calculate the interference between waves starting from a point and arriving at another point by adding together all possible connecting paths. The interference between rays on different paths will be destructive for all these paths except the "classical" path described by Snell's law, where it is constructive. (This is a simplified account—in fact the optical path for light is an extremum, and in some cases may be a maximum.)

The study of refractive index subsequently became central to the design of optical instruments, lenses, cameras, and eye-glasses, all of which must bend the direction of a light ray toward a focus. The detailed study of the atomic processes responsible for variations in refractive index has led to many other useful effects, such as Polaroid sunglasses, birefringence, anti-reflection coatings on cameras, phase-contrast microscopy, and the total internal reflection effect. This causes the surface of water to appear as a mirror when viewed upwards from below by fish or divers, outside a small transparent circular window directly overhead. In recent exciting developments, new "metamaterials" have been discovered in which the refractive index is negative—these are the materials used for "cloaking" experiments, in which objects can be made invisible by bending light rays around the object, like Harry Potter's invisibility cloak.

But none of this early work provided a value for the speed of light, and Fermat's original ideas had many critics. The first reasonably accurate measurements of the speed of light came from astronomy, from the outstanding Danish astronomer Ole Roemer, in one of the greatest experiments in the history of physics.

2

Ole Roemer, Who Started It All

Ole Roemer was born on September 25, 1644 at Aarhus in Denmark to a family of merchants (**Figure 2.1**). Following classes at the local cathedral school he attended Copenhagen University from 1662, enrolling in medicine. However he was befriended by the Professor of Mathematics, Dr Erasmus Bartholin (1625–98), the discoverer of birefringence in the mineral iceland spar, who introduced him to problems in astronomy. Roemer lived with the Bartholin family, learning from Bartholin both astronomy and mathematics. Bartholin was impressed by his abilities, and entrusted him with the job of editing the observations in the manuscript of Tycho Brahe.

At the time, the problem of the determination of longitude for ships at sea occupied many of the best scientists in Europe. Galileo had proposed that observation of the orbital motion of the moons of Jupiter by telescope could help, by providing a kind of universal clock. He had discovered four moons with his first telescope in 1610—today we know there are at least sixteen.

Lines of longitude run from the North Pole to South Pole in equally spaced angular increments from zero degrees at Greenwich UK to 360 degrees after circling the globe. Since the earth rotates by 360 degrees every twenty-four hours (one degree every four minutes), one's longitude, or angular distance around the Earth from Greenwich, may be found if the time in Greenwich (where the longitude is arbitrarily defined to be zero) is known when the local time (where you are) is exactly noon. The establishment of long-wave global radio transmission after 1900 by Marconi immediately solved the longitude problem once and for all—one of the first things done with radio was to transmit the time at Greenwich to all points of the globe, immediately giving everyone their longitude, by comparison with their local noon. Local noon can be identified from the angular height of the sun above the horizon, which reaches its maximum at noon. On the equator, for example, if local noon occurred at exactly midnight Greenwich time,

Lightspeed: The ghostly Aether and the race to measure the speed of light. John C. H. Spence.
© John C. H. Spence 2020. Published in 2020 by Oxford University Press.
DOI: 10.1093/oso/9780198841968.001.0001

Figure 2.1 Ole Roemer. (From Google Images)

you would have to be located directly opposite Greenwich, on a line through the center of the Earth, which would emerge on the international date line north of the tip of New Zealand. Before the advent of radio, accurate mechanical chronometers set to Greenwich time were taken on voyages. Harrison's famous chronometer, the first to be sufficiently accurate for longitude determination, with its torsion pendulum (immune to vibration) and temperature compensation, did not finally appear until 1761. Latitude was simpler—it had been readily available for centuries from the angular height of the pole star (in the northern hemisphere) or the height of the sun above the horizon at noon.

The most suitable of Jupiter's moons is the innermost, Io, which orbits Jupiter every 42.5 hours, so that the eclipse of Io, as it disappears behind Jupiter, can be used for regular time keeping. The plane of Io's orbit is similar to that of Jupiter's orbit around the Sun. This would be much more accurate than a mechanical clock, such as a pendulum clock subject to the violent motion of a ship in a storm. Unfortunately it proved impractical to observe Io from a rocking ship by telescope, but did provide a workable method on land, allowing longitude determination throughout the world.

The use of these Jovian eclipses for navigation required a table giving their time of occurrence at some reference point on Earth, such as Paris or Greenwich. The Italian astronomer Giovanni Cassini (1625–1712) had taken up Galileo's proposal to use observations of Io for longitude determination, and published such a table. In 1671 Cassini, who had moved from Bologna to Paris to head the Paris Observatory in 1668, sent Jean Picard to Bartholin in Copenhagen to test the longitude method by observing the Io eclipses from the old observatory of Tycho Brahe (1546–1601), at Uraniborg on the island of Hven near Copenhagen, whose longitude was known approximately, and to obtain its location more accurately. We can think of the regular Io eclipses as a clock, ticking every 42.5 hours and used to calibrate or reset a local clock taken on travel, which gives the time back in Paris, whose longitude might be defined as zero degrees (as an alternative to Greenwich). If Picard saw the Io eclipse later in Uraniborg than Cassini did in Paris, the longitude difference between Paris and Uraniborg could be calculated by proportions, knowing that it takes twenty-four hours for a full rotation of the Earth. This could then be compared with the currently accepted value for the longitude of Uraniborg. The time difference they observed was about forty-two minutes, being the time between when the Sun would appear directly overhead in Paris and when it would do so in Uraniborg, due to the rotation of the Earth. (Recall that a time difference of twelve hours would put Uraniborg near New Zealand.)

Bartholin and Picard took the young Roemer with them to make the observations in the ruins of Brahe's observatory. Impressed by his ability, Picard invited Roemer to return to Paris to work on that data and other data which Cassini had collected at the Paris observatory as its director. There Roemer was given the job of teaching mathematics to the eldest son of the French King, the heir to the throne, and was able to live in the observatory. His inventive skill in design and instrumentation became clear as he constructed a greatly improved micrometer for measuring very small astronomical angles, and invented a famous planisphere to provide a model of planetary motions. He was also involved in the construction of the fountains at Versailles.

Cassini, as head of the Paris observatory, had tasked his nephew Maraldi and Roemer with analyzing the orbits of the Jovian moons. Seasonal variations in the period of the orbits were noticed, depending on when the measurements were recorded, and these were attributed to various causes, for example, elliptical rather than circular orbits. However these discrepancies provoked a far more important and novel idea, that the variations might be due to the finite speed of light having to travel different distances to the Earth between eclipses as the Earth orbits the Sun. This idea was initially written down by Cassini (possibly suggested by Roemer), however he quickly refuted and steadfastly opposed this idea ever after, once a full analysis had been presented by Roemer.

In September 1676 in an address to the Paris Academy of Sciences, Roemer predicted the late arrival (by ten minutes) of an eclipse of Io on November 9, due to the finite velocity of light, adding that it would take light about eleven minutes to travel from the Sun to the Earth (modern measurements give eight minutes and nineteen seconds, about 500 seconds in total). The prediction was based on a new *equation of light*, and used about thirty eclipse observations. His prediction was confirmed, bringing him immediate recognition and allowing him to publish his analysis in *Journal de Savants* in November 1676. (A translation of the paper in English was published in *Philosophical Transactions* in 1677, and is given in Appendix 2).

The delay was obviously important, since it would lead to errors in longitude determination. Measurements of the moment of eclipse at that time were accurate to about thirty seconds. He realized that the delay would accumulate over many Io orbits, leading to longer periods of measurement and increased accuracy. On reading the *Philosophical Transactions* paper, Huygens wrote to Roemer requesting more detail. Roemer's reply completely convinced Huygens of the accuracy of his analysis, and Roemer was then able to read out Huygens's subsequent reply at an Academy meeting in support of his ideas against the growing opposition of his boss Cassini, who now strenuously opposed the idea of a finite light speed. It was Huygens who first combined Roemer's estimate of the time delay with the speed of the Earth in orbit to obtain the first value for the speed of light in 1690. The arguments against Roemer's analysis, attributing the delay to various sources of error, including eccentricity in Io's orbit, were published in a lengthy paper by Maraldi in 1707.

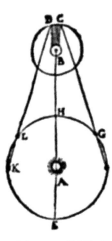

Figure 2.2 Roemer's original diagram (1676). The Sun at A, orbited by the Earth at K, with Jupiter at B, and its moon Io at C. (From Cohen (1942).)

Figure 2.2 shows the diagram Roemer drew in his paper. The sun is at A, and the Earth moves between G, H, L, K, and E around the Sun throughout the year. Jupiter is at B, and its moon Io moves from C to D, being in the shadow of Jupiter (as seen from the Earth) between these points.

If the Earth were stationary at L (or moving around H), an observer on the Earth would measure a time interval of 42.5 hours between eclipses, the times at which Io first emerges from behind Jupiter at D. But if the Earth moves from L to K while Io is performing its orbit, the measured time between eclipses will be longer by the time (about fifteen seconds) it takes light to travel from L to K to catch up with the Earth. The Earth moves at about 30 Km per second or 18.6 miles per second relative to the Sun, and so moves about 2.8 million miles between eclipses. On the other hand, if, six months later, the time between eclipses is measured while the Earth is moving toward G in an anticlockwise direction, light will have less distance to travel and this time will be shorter. This was Roemer's explanation for the variations in orbital periods of Io found among many observations using the telescopes recently invented by Galileo. His explanation gave the strongest evidence to date that light does not travel instantaneously, and if the diameter of the Earth's orbit around the Sun were known, it could be used to estimate the speed of light with reasonable accuracy for the first time.

The diameter of the Earth's orbit was not accurately known at that time, although in 1672, as we shall see in a later chapter, Cassini himself used the method of parallax to obtain an approximate radius for the Earth's orbit of eighty-seven million miles, not far from the currently accepted value of ninety-three million miles (equal to one astronomical unit of distance). Kepler's laws, one of which relates the ratios of the known periods of the planets to the ratios of their orbital radii, were well known at the time, so that a measurement of one planetary radius would fix them all. This value of Cassini's had a large error associated with it, but was accepted and widely used for a century, before being improved to within three percent of the current value using the 1769 "transit of Venus" observations, in which James Cook took part during his first voyage of exploration to Australia. When combined with Roemer's time measurements, Cassini's radius gives a value of about 214,000 km/s for the speed of light, compared with the modern value of 300,000 km/s or 186,000 miles per second. But Roemer had achieved his most important aim—to show that the speed of light was finite. He himself does not appear to have given a value for the speed in familiar units, perhaps because of the large errors in distance measurements—this was left to Huygens in 1690.

If D is the distance between the Earth's position at the first and last of n eclipse observations, and the true period of Io (measured at H) is T, then the speed of light c is

$$c = D / (t - n T),$$

where t is the time between first and last eclipse observations. The simplest, but impractical, case occurs when D is the diameter of the Earth's orbit, and eclipses are observed over six months between H and E in Figure 2.2. The true period T of Io can also be obtained more accurately by dividing one year by the number of orbits counted in a year, for which the advances and retardations cancel. If the Earth is moving away from Jupiter at speed v, then the apparent period of its moon will be $T' = T (1 + v/c)$, where T is the true period. The same result can be used to find the relative velocity v of any galaxy emitting light, where T now becomes the period of the light waves emitted by some hot atoms, whose true period can be measured using these same atoms in a laboratory on Earth.

Cassini never accepted Roemer's result, but it was accepted by Newton, whose book *Opticks* in 1704 gives "seven or eight minutes" for light to travel from Sun to Earth (crediting Roemer), a value closer to today's value than Roemer's. It was also accepted by Huygens, Leibnitz, Flamsteed (the British Astronomer Royal), and the great British astronomers Edmond Halley and James Bradley. Critics attributed the timing delays to eccentricities in Io's orbit, and it is noteworthy that good measurements to confirm Roemer's ideas could not be obtained from the other moons of Jupiter, a fact which appears to have caused Roemer himself to have transient doubts about his theory. But Bradley's independent measurement by a different method, which we discuss in a later chapter, was published in January 1729, and this agreed with Roemer, convincing even the French at last. Even the skeptical Maraldi published an analysis based on Roemer's equation of light for the motion of the third satellite of Jupiter in 1741. In 1817, J. Delambre published a summary of over a thousand observations of Jupiter's satellites, from which he obtained an estimate of 493.5 seconds for the time it takes light to travel from the Sun to the Earth. (The modern value is very close to 500 seconds, or eight minutes and twenty seconds.) So the most important error in determinations of the speed of light lay with estimates of astronomical distances, as we discuss in Chapter 3.

In 1679, Roemer went to England where he met Newton, Halley, and Flamsteed and visited the Royal Society to examine their *seconds pendulum*, used to define a unit of time. Roemer read many papers to the French Academy and has been credited with work on refraction and measurement of the speed of sound. But Roemer, a Protestant in Catholic France, was eventually forced to leave Paris in 1681, due to prejudice against his religion. He took up a position as Professor of Mathematics at Copenhagen University, and soon after was appointed Astronomer Royal by King Christian V. In 1699 he became, nevertheless, a Foreign Member of the French Academy of Science. His Danish appointment led rapidly to a host of other appointments, including Master of the Mint, Inspector of Naval Architecture, and member of the King's Privy Council. The next Danish King, Frederick IV, clearly impressed by his abilities, made him a senator and head of the State Council. In 1693 he became a chief justice of the Supreme Court of Copenhagen, and later was responsible for reform of the tax system. He became Mayor of Copenhagen in 1705, and Prefect of Police soon after, in addition.

Roemer devoted a lot of time to attempts to reform the calendar, urging unsuccessfully for Denmark to adopt the Gregorian calendar used elsewhere. His interest in instruments and hydraulics led him to the discovery of the best shape for the teeth of gears to minimize friction, the epicycloid. He sent this finding to Huygens, suggesting that he use it in his clocks. He also devoted a lot of effort to the design of a better thermometer, being the first to realize that two fixed points were needed to calibrate a temperature scale. In a letter, Daniel Fahrenheit (1686–1736) writes that on a visit to Denmark he met Roemer, who first showed him that two reference points were needed to fix a temperature scale, which Roemer had taken as the melting point of a mixture of snow and sal ammoniac, and the boiling point of water. From this, it is clear that Roemer invented the thermometer. At his death on September 19, 1710, fifty-four instruments of his invention were inventoried for his estate.

Unfortunately, most of the vast accumulation of notes by him on his observations were destroyed in a fire in 1728, but a few were saved during the fire by his devoted accomplice Peder Horrebow, as vividly described in his book. Roemer also sent observations to friends which have survived, and maintained a "commonplace book" for notes, entitled *Adversaria*, which he kept by his window at the library of the University of Copenhagen, and which was discovered there early in the last century, and has since been published. Horrebow's 1735 book (still in print) contains much detail (in Latin) on Roemer's methods, including the picture of his transit telescope, shown in **Figure 2.3**, which Roemer installed in an observatory built at his own expense in 1704 near Vridloesemagle (between Copenhagen and Roeskilde). The telescope could be aligned perpendicular to the horizontal rotation axis by reversing it. Note the system of counterweights and the pendulum clock, ticking loudly to mark the instant of transit. Observations were made in the plane of the meridian (a circle around the Earth of constant longitude) or in the prime vertical. The instrument had crosshairs at the common focus of the eyepiece and objective lenses (as in any modern design) illuminated from the side, a remarkable advance for the time. In 1700, Leibnitz wrote to Roemer, asking for advice on the design of a good observatory. His reply is full of detail of his Vridloesemagle facility, and his choice of dimensions, intended to improve on his experience at the Paris Observatory.

As a result of Roemer's work, tables giving the corrected times of Io's eclipses (Ephemerides) were published and used for longitude determination on land until about 1800. At sea, Harrison's chronometer became popular from about 1750, and by 1800, longitude could be determined

Figure 2.3 Roemer's transit telescope. Note pendulum clock and counter-weights. From Horrebow (1735).

in this way within a fraction of a degree on trips of several months' duration.

Roemer died on September 19, 1710 of "tortor calculus." He had married twice, first to Erasmus Bartholin's daughter Anne Marie and, after her death in 1694, to Elisa Bartholin in 1698, who outlived him. A modest and generous man, too busy to promote himself, but nevertheless clearly a recognized scientific leader of his time, he was unlucky to have his scientific legacy incinerated soon after his death, with the result that he is remembered today only for being the first to measure the speed of light, despite his many other accomplishments.

3

Measuring the Cosmos
Parallax and the Transit of Venus

To measure the speed of light, how far it travels in a given time, we need to measure that distance. As we have seen those distances must be very large indeed if the time is to be easily measured using the pendulum clocks of past centuries; such as the eleven minutes Roemer estimated for light to travel the ninety-three million miles from the Sun to Earth. The story of mankind's attempts to measure astronomical distances, going back to the time of the ancient Greeks, has filled volumes, and we now sketch the main highlights.

The method used prior to the twentieth century was based on the parallax effect, which became an abiding obsession of astronomers for centuries because the observation of a parallax effect would confirm Copernicus's heliocentric idea that the Earth orbits the Sun. This method now gives good estimates of planetary orbits and distances to nearby stars. Kepler's law, relating orbital radii to the known orbital periods, then allows us to find all the planetary radii once one of them is known. His 1619 law states that the squares of the orbital periods of the planets are proportional to the cube of their mean radii. **Figure 3.1** shows the principle of parallax, which will be familiar from the way in which a finger, held in front of your eye, appears to move sideways with respect to a distant wall if the head is moved sideways. To use this effect to measure the distance to a star, we must distinguish those stars in a very distant group which provide a fixed background screen, from a nearby planet or star shown at C. (The planets—from the Greek word for *wanderer*—orbiting the Sun can be distinguished from stars by their rapid motion.) Then, provided that the diameter of the Earth's orbit D is known, the star C will appear to have moved across the background on the right when viewed from A, compared with its position six months later, when the Earth is at B.

Lightspeed: The ghostly Aether and the race to measure the speed of light. John C. H. Spence.
© John C. H. Spence 2020. Published in 2020 by Oxford University Press.
DOI: 10.1093/oso/9780198841968.001.0001

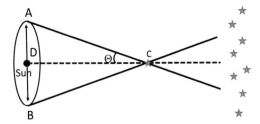

Figure 3.1 The principle of parallax used to measure stellar distances from Earth. The Earth orbits the Sun around the circle AB with diameter D. A planet is shown at C against a background of fixed stars at the right. A different group of background stars will be seen behind the planet if we observe first from A, then six months later from B.

The distance D from A to B is the baseline, and if the angle Θ can be measured for a known baseline distance, simple trigonometry gives the distance between the Sun (or Earth) and the star at C. Baseline separation is also used to judge distances in rangefinder cameras, and by our eyes, which provide simultaneous views from two slightly different directions. The brain compares these different views, allowing us to estimate distance. Unfortunately, for the stars, this angle Θ is extremely small, about half an arcsecond for the nearest stars. (An arcminute is one sixtieth of a degree, one arcsecond is a sixtieth of an arcminute, written 1″). The inability of astronomers to measure this tiny angle, due to instrumental limitations, was used by critics to deny the validity of the Copernican model of the Earth orbiting the Sun. For our nearest star, the small red dwarf Proxima Centauri (discovered in 1915) at 4.2 light years from our Sun, the parallax is 0.76 arcseconds, equivalent to viewing a one-inch object from a distance of 4.2 miles. It is no wonder that it was the invention of the telescope in the sixteenth century which was needed to make parallax measurements practical.

Except for the Moon. We see the Moon against a background of stars, and if this is done at the same time from two places a few thousand miles apart, the Moon will appear to shift by a distance equal to its diameter against the background of the fixed stars, which is easily noticeable without a telescope. Knowing the baseline, the distance between the viewing places on Earth, and the angular shift in the Moon's position, the distance from Earth to the Moon can easily be found. It is remarkable that the ancients did not think of this, in spite of the difficulty of making observations at two places at the same time at night thousands of miles apart without a camera.

In 350 BC Aristotle had proposed a spherical Earth (suggested earlier by Pythagoras), based on the evidence of the curved shadow of the Earth seen on the Moon during a lunar eclipse, perhaps also on the disappearance of ships over the horizon, and the changing appearance of the stars during a long journey. Some eighty years later, Aristarchus of Samos appears to be the first to suggest that the Earth orbits the Sun, and his book (which still exists) "*On the Sizes and Distances of the Sun and Moon*" gives estimates of these quantities, placing the planets in the correct order of distance from the Sun. **Figure 3.2** shows a page from his book.

Figure 3.3(a) shows his correct argument that at half-moon, one has a right-angled triangle in which the angle between the Sun and the Moon can be measured from Earth, giving the ratio of Earth–Moon to Earth–Sun distances. He measured this angle to be 87°, obtaining a distance to the Sun from Earth of between nineteen and twenty times the distance from Earth to the Moon, with S/M ~ 19.1. (The angle should actually be 89.8°, giving 390 times.) His estimate was accepted for about 2000 years until the invention of the telescope.

Aristarchus used three other constructions to obtain further information. He noticed that during an eclipse of the Sun, the Sun's disk is almost exactly filled by the Moon, as shown in Figure **3.3(b)**, a lucky coincidence. From this it follows that the two triangles formed between

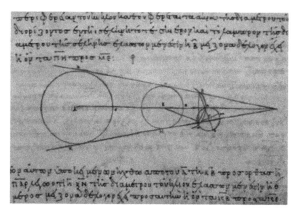

Figure 3.2   Aristarchus's calculation of the relative sizes of the Earth, Sun, and Moon. From a tenth century Greek copy. (From Wikipedia.)

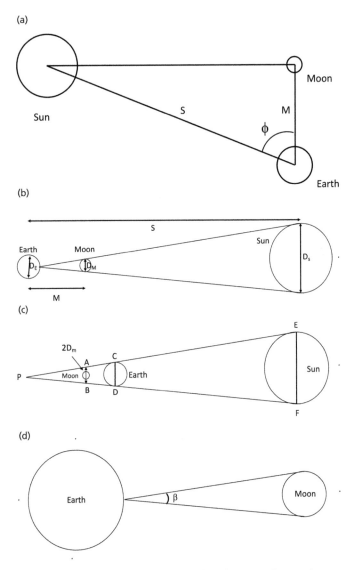

Figure 3.3 (a) Trigonometry gave Aristarchus the ratio of S to M by measuring the angle φ at half-moon. (b) Aristarchus's second construction. (c) Aristarchus's third construction. (d) Aristarchus's fourth construction.

the Earth and the Moon, and between the Earth and the Sun, are similar triangles. Euclid's theorems on geometry had shown that in this case

$$D_s / D_m = S/M.$$

So from his previous value of S/M he now had the ratio of the size of the Sun to that of the Moon as about twenty (it should be 390). In Figure 3.3(c), Aristarchus studied an eclipse of the Moon, when the Moon passes on the far side of the Earth across the shadow of the Earth. By timing the progress of this shadow across the Moon's surface, he deduced that if the Moon were twice as big, it would exactly fill the Earth's shadow, which converges to the point P. Now we have three similar triangles, PAB, PCD, and PEF, whose sides are in corresponding ratios. Using these expressions and the previous result, we obtain after some algebraic manipulation

$$D_e = 3 D_m D_s / \left(D_s + D_m \right)$$

for the diameter of the Earth. Using the earlier result for D_s/D_m, we obtain $D_e/D_m \sim 2.85$ from this result for the ratio of the size of the Earth to that of the Moon, which should actually be 3.67. He could also get a value for the ratio of the size of the Sun to that of the Earth, obtaining $D_s/D_e \sim 6.7$ (it should be about 109). Steven Weinberg, in his fascinating book on the history of astronomy, explains why Ptolemy's system, although incorrect, worked so well. Weinberg also indicates that this result of Aristarchus, that the Sun is much bigger than the Earth, was crucial in convincing Aristarchus that the Earth orbits the Sun, rather than the Sun orbiting the Earth.

A final observation, shown in **Figure 3.3(d)**, was simply his estimate of the angle subtended by the Moon at the Earth, $\beta = D_m/M$ (in radian angular measure), as shown, which he estimated to be two degrees (it should be about half a degree, the value given correctly by Archimedes). Putting all these results together, Aristarchus could estimate the ratio of the distance of the Moon to the diameter of the Earth, and the ratio of the distance to the Sun to the diameter of the Earth (for which he obtained 191 instead of the correct value of 11,600). Two other geometric constructions have been used, the lunar parallax triangle formed between the center of the Earth, an observer on the surface of the Earth, and the Sun, and the diurnal parallax method, which uses the daily rotation of the Earth to sweep an observer across the baseline. The apparent movement of a nearby star against the background of fixed stars is observed between the time when the star rises above the horizon (so that its rays arrive tangentially to the surface of the Earth, just brushing the surface), and about six hours later when it is directly overhead. One then has a right-angle triangle formed by the center of the Earth, the observer, and the star, with the baseline being the radius of the Earth.

In a geocentric universe, the widest baseline possible is the width of the Earth; if Artistarchus were correct, and the Earth orbited the Sun, astronomers could use the much larger Earth's orbit around the Sun (AB in **Figure 3.1**) for greater accuracy, which they were later to do.

The actual distances between Earth, Moon, and Sun, rather than their ratios, would not be known until Eratosthenes later measured the actual size of the Earth, from which all these other distances could then be deduced using Aristarchus's ratios. Born in 276 BC in present-day Libya, he trained as a Stoic philosopher in Athens, moving on to the Platonic Academy. He was a talented poet, mathematician, and geographer, an all-rounder, and friend of Archimedes, perhaps most famous for his method of finding prime numbers, the *sieve of Eratosthenes*. He earned the nickname *Beta*, for being second-best at many things. In 245 BC he became librarian at the Library of Alexandria, the cultural center of the Greek world at that time, where he remained for the rest of his life, and was later appointed head of that great library by the King of Egypt. The library was the first known to arrange books alphabetically by author, and became the most important intellectual center for scholars in the western world. Eratosthenes opposed the conventional idea of that time that the peoples of the world could be divided into two groups, Greeks and Barbarians, instead arguing that there was good and bad among all peoples. He wrote two geography books, one of which was the first to provide a map of the known world on a grid of lines of latitude and longitude. On becoming blind, he committed suicide in 194 BC.

Eratosthenes used the length of the shadow of a vertical stick at Syene, about 500 miles away, when it was exactly noon at his home in Alexandria, to estimate the size of the Earth. This stick could be the vertical piece on a sundial, known as a *gnomon*. The idea for the experiment may have come from someone noticing that the shadows from identical sundials have different lengths at different places at the same time. Local noon could be judged from the sundial or by looking at the

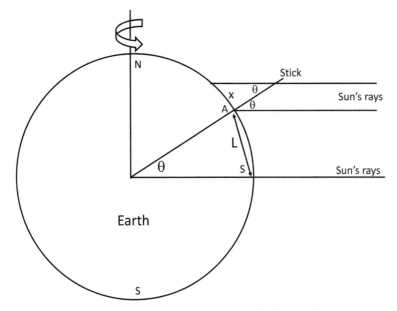

Figure 3.4 Eratosthenes, at around 200 BC measured the angle θ by measuring the length x of the shadow of a stick at A (Alexandria), when there was no shadow at S (Syene), from which he found the circumference of the Earth, knowing L.

image of the Sun on water down a deep well. His method is shown in **Figure 3.4**, where the length of the shadow x gives the angle θ, which he found to be 7.2° in Alexandria at A when the Sun was directly over-head at noon in Syene S (on Elephantine Island, today near Aswan). Because the Sun looks so small, and is so far away, its rays arrive parallel to each other at Syene and Alexandria. From this angle and the distance between Syene and Alexandria he could use simple proportions to find the circumference C of the Earth. Since there are 360° in a circle, we have θ/360 = L/C, where L is the known distance between Syene and Alexandria. He found this ratio to be about 50 (it is actually 47.9). Presumably L was measured by trained walkers counting their steps— there is a debate among historians as to the length of the distance unit used at that time (the stadion), but the possible values don't vary much. Eratosthenes had been told that at the summer solstice (the longest day), the Sun was directly overhead at noon in Syene (zero shadow length), which told him when to measure the shadow in Alexandria.

He took Syene to lie due south of Alexandria (on the same meridian), so the Sun would be at maximum height at noon in both places at the same time (but only exactly overhead at Syene). Syene lies on the Tropic of Cancer, where the sun is exactly overhead only once a year on the longest day, at which moment it casts few shadows—none from a vertical stick. (North of the Tropic of Cancer the Sun is never directly overhead, South of it, but north of the equator, the Sun is directly overhead twice a year). From this circumference C, fifty times the distance from Alexandria to Syene, he could find the radius R of the Earth, using Euclid's result that $R = C/2\pi$. Remarkably, Eratosthenes obtained a value within 15% of the correct value of 3959 miles. A recent repeat of this experiment using modern measurements gives an error of 0.16%. His estimate was used for thousands of years—its importance lay in the fact that now geographers could calculate the distance between locations of known latitude, while it also provided a unit of distance which could be used in further measurements of the size of the solar system—how many Earth diameters to the Sun? It is fun to repeat his experiment using ordinary maps on the web—just find two cities a known distance L apart on about the same meridian (line of longitude) and find the difference in latitude θ between them, from which you can find the radius of the Earth, as above.

Hipparchus, in around 120 BC, was perhaps the greatest of the ancient astronomers, and made extensive use of trigonometry to provide the first trigonometric tables, and to make heliocentric models of the Earth's and the Moon's motions based on his and earlier observations by the Babylonians, Aristarchus, Eratosthenes, and others. His accurate quantitative predictions for the motions of the Sun and Moon have survived to the present day. He also estimated Earth and Sun sizes and distances. His estimate of the Earth–Moon distance (based on eclipse observations at two different places) gave between seventy-one and eighty-three Earth radii (it should be sixty). Most of his original fourteen books have been lost, but we know of his work from later reports by others, and his famous star catalog provided the basis for Ptolemy's catalog, from which Hipparchus's has been reconstructed. There is evidence that he may have predicted solar eclipses and discovered the wobble (precession) in the Earth's rotation, and is thought to have been the first to calculate circular orbits for the planets around the Sun. As the most authoritative astronomer of Greek antiquity, his opinions dominated western thought for 2000 years until the time of Copernicus.

A crater on the Moon and an astronomy satellite have been named after him.

All these astronomers greatly underestimated the Earth–Sun distance, because they used parallax arguments based on comparisons of the apparent sizes of the Sun and Moon, and the size of the Earth, but were limited to about three degrees (10,800 arc seconds) in accuracy by the ability of the unaided eye to judge angles.

In summary, the Greeks used both geometrical constructions and the ideas collected in Euclid's *Elements*—what we would now call trigonometry—to make various estimates of the size of the Earth, Earth's orbit around the Sun, and the distance to the Sun and Moon. Their methods were ingenious and sound, but errors in experimental measurements produced gross errors in their results.

Ptolemy (at around AD 100–170), the next of the great astronomers, in his Almagest, produced a star catalog and tables to predict future positions of the planets, this becoming the authoritative text on astronomy for more than a thousand years, until the time of Copernicus. His geocentric model (with Sun and planets orbiting the Earth) gave a gross underestimate of 1210 Earth radii for the average distance between the stationary Earth and the orbiting Sun, with fixed stars all located on a sphere of radius 20,000 times that of the Earth. In spite of all these errors, his tables, star calendar, and almanac were able to predict the positions of the Sun, Moon, and planets and the eclipses of the Sun and Moon. He also produced important treatise on geography, music, astrology, and optics.

Copernicus's great text *On the Revolutions of the Heavenly Spheres*, like Aristarchus, put the Sun rather than the Earth at the center of the universe. It was published in 1543, the year the Polish astronomer died. His great successors were Tycho Brahe (1546–1601) and Johannes Kepler (1571–1630), who collaborated with Tycho in Prague. Brahe believed that the Sun orbits the Earth, but all the planets except the Earth orbit the Sun. In an intellectual climate of strict Aristotelian philosophy, only mathematical simplicity recommended Copernicus's heliocentric theory to astronomers, which placed the Sun at the center of the universe. It was Kepler and Galileo, fully accepting Copernicus's ideas fifty years later, and finally the unifying principle of Newton's law of gravity (1687), which led to general acceptance of the heliocentric model.

In the early nineteenth century, with reasonably good estimates of the time taken for sunlight to reach Earth, getting an accurate value of the speed of light required more precise estimates of the Sun–Earth distance. The Earth's diameter was fairly accurately known. Bessel, in 1841, gave a value very close to the modern value. The Earth–Sun distance could therefore be obtained from the solar parallax, that is, the angular size of the Earth as seen from the Sun, when combined with the known size of the Earth. The solar parallax could be estimated from a transit of Venus, when Venus passes across the face of the Sun, and the Earth–Sun distance (one Astronomical Unit, or AU) then estimated using a method first proposed by James Gregory in 1663. Kepler was the first to predict a transit of Venus (for 1631), but his prediction was not accurate enough to determine the places on Earth from which it would be visible.

Venus has an almost perfectly circular orbit, with an orbital year of 225 days. At transit, Venus passes between the Sun and the Earth, crossing the Sun's circular image as a small black disk. The transits occur in pairs eight years apart about every century (the last one was in 2012). The Sun's bright image can be projected onto a screen in a darkened room using a telescope, and its angular diameter is easily measured from the angle subtended by the size of this disk image at the telescope lens. This image of the Sun can also be used as the backdrop (like the fixed stars in Figure 3.1), as Venus moves across the Sun's disk during transit. **Figure 3.5** shows the situation. The black disk could be viewed from opposite sides of the Earth by two observers at the same time, and

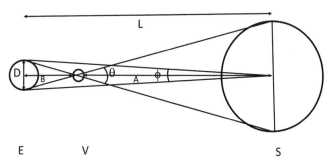

Figure 3.5 The Earth (E), Venus (V), and the Sun (S), showing the solar parallax angle ϕ and the Venus parallax angle θ. Both can be measured from observations of the transit of Venus across the Sun from two different places on Earth at the same time. Knowing that $A/(A + B) = 0.72$, from Kepler's work, it is then possible to work out the Earth–Sun distance L using the Earth's known diameter D as a baseline.

their images of the Sun compared. In reality, the path of the small image of Venus moving across the Sun's disk is tracked in detail by several observers around the world with synchronized clocks (not a simple matter in past centuries!). These images will show Venus projected onto different places on the Sun's surface due to the parallax effect, between which the angle Θ can be measured from Earth. Kepler's third law, using the known period of the Earth and Venus's solar orbits, gives Venus's distance from the sun as 0.72 times that of the Earth (Mars is 1.5). Thus, in the figure, $A/(A + B) = 0.72$. From the known distance D between the observers on Earth, and measurement of the Venus parallax angle θ, we can find B by trigonometry, giving A and hence the Earth–Sun distance $L = A + B$ and the solar parallax angle ϕ.

Jeremiah Horrocks, following Kepler's predictions of transits for 1631 and perhaps 1639, observed the 1639 transit on November 24 by focusing an image of the (moving) Sun onto a sheet of paper on an interior wall of his house, and tracing the result as Venus crossed the Sun's disk over a period of a few hours. Clouds passing over the Sun threatened all this early work. His companion Crabtree, at another location, estimated the angular size of Venus to be 1' 3", within one second of the arc of the modern value. In his report *Venus in Sole Visa*, Horrocks wrote:

> *When the time of the observation approached, I retired to my apartment, and having closed the windows against the light, I directed my telescope, previously adjusted to a focus, through the aperture towards the Sun and received his rays at right angles upon the paper . . . I watched carefully on the 24th from sunrise to nine o'clock, and from a little before ten until noon, and at one in the afternoon, being called away in the intervals by business of the highest importance which, for these ornamental pursuits, I could not with propriety neglect . . . About fifteen minutes past three in the afternoon, when I was again at liberty to continue my labors, the clouds, as if by divine interposition, were entirely dispersed . . . I then beheld a most agreeable spectacle, the object of my sanguine wishes, a spot of unusual magnitude and of a perfectly circular shape, which had already fully entered upon the Sun's disk on the left . . . Not doubting that this was really the shadow of the planet, I immediately applied myself sedulously to observe it . . . although Venus continued on the disk for several hours, she was not visible to me longer than half-an-hour, on account of [the Sun] so quickly setting . . . The inclination was the only point upon which I failed to attain the utmost precision; for, owing to the rapid motion of the Sun, it was difficult to observe with certainty to a single degree . . . But all the rest is sufficiently accurate, and as exact as I could desire.*

Based on these observations and Kepler's work, Horrocks estimated the Earth–Sun distance to be about sixty million miles, the most accurate estimate to that time. The correct distance is about ninety million miles. Horrocks died at the age of twenty-two and is seen by many as a founder of British research astronomy. The house where he worked and probably made the observations, Carr house, in the village of Hoole near Preston in the UK, still stands, and displays an historical plaque. Even the window he probably used has been identified.

Edmund Halley (1656–1742), in 1716, suggested that the 1761 and 1769 transits of Venus might be used to estimate the Earth–Sun distance, based on an earlier method proposed by James Gregory, but he died before it could be done. Halley, the second Astronomer Royal (after Flamsteed) was a powerful force in eighteenth-century science. In 1676 he set up an observatory on the island of Saint Helena in the South Atlantic to catalog the stars of the southern hemisphere for the first time. This was later published in 1679 with a few hundred new stars, resulting in him being elected a Fellow of the Royal Society. During his time on Saint Helena, he observed the planet Mercury passing over the face of the Sun (a transit), and realized that a transit of Venus could be used to determine the size of the solar system. He had wide-ranging interests in science—he established the relationship between barometric pressure and altitude, and correctly attributed atmospheric motion to solar heating. Like John Flamsteed before him, he became obsessed with the problem of deriving Kepler's laws from first principles, and in 1868, went up to Cambridge to ask Newton about this. According to a later reminiscence by a friend, Halley asked Newton "what curve would be described by the planets, supposing that the force of attraction to the Sun be reciprocal to the square of their distance from it?" Newton said he had already solved the problem, but could not find his analysis, and would send it to him when he did. Eventually Newton's *On the Motion of Bodies in an Orbit* arrived, in which the elliptical orbits of planets and comets were shown to result from gravitational attraction to the Sun according to an inverse-square force law, while Kepler's laws were properties of these orbits. Newton later acknowledged that he had been stimulated to undertake this work by correspondence with Robert Hooke in 1679—Hooke's contribution has been the source of much scholarly debate ever since. Realizing the importance of Newton's work, one of the greatest discoveries in the history of science, Halley returned to Cambridge to urge Newton to publish it. Instead, Newton expanded it to become his greatest work, his *Principles of Natural Philosophy and Mathematics*, which was published at Halley's expense in 1687.

On the basis of Newton's theory, Halley was then able to predict the motion of the comet Kirch, but unfortunately got the period wrong. Later, in 1705, he found he could fit Newton's orbit to comet observations recorded in 1456, 1531, 1607, and 1682, suggesting that they all came from the same comet, which he then predicted would return in 1758. The confirmation of this prediction, which Halley did not live to see, was a triumphant confirmation of Newton's theory of gravitation. That comet came to be known as Halley's comet, perhaps the most famous of them all. It revisits the Earth about every seventy-five years, when it can be seen with the naked eye, most recently in 1986. So it would be possible to see it twice in a lifetime. The 1986 appearance was closely studied by spacecraft, confirming that the comet was a "dirty snowball," a mixture of volatile ices and dust with a non-volatile surface.

Halley had been rejected from faculty positions at Oxford on the grounds of his open-minded approach to theological questions (he was not an atheist), but following the death of his theological enemies, in 1703, he became Professor of Geometry. He worked on many other problems throughout his life, learnt Arabic and Greek in order to translate Apollonius's *Conics*, worked on the dating of Stonehenge, and discovered the "proper motion" of the fixed stars. This is the motion of stars relative to the center of mass of the solar system (a point near the center of the Sun), and compared with the background of the most distant stars. It can be used to determine the motion of a star relative to our Milky Way galaxy. Halley made exploratory voyages (as Captain of the *Paramour*), mapping compass variations over a wide portion of the Atlantic Ocean. His crew found him incompetent, leading to an inquiry. As Andrea Wulf describes in her book, he was said to "talk, swear, and drink brandy like a sea-captain."

In 1716, Halley proposed an accurate measurement of the Earth–Sun distance by timing the transit of Venus expected in 1761 and 1769, using the method proposed in 1663 for the transit of Mercury by James Gregory, the inventor of the Gregorian telescope. Halley almost certainly had a copy of Gregory's book, but never acknowledged his use of Gregory's method, which is explained in **Figure 3.5** and also shown in **Figure 3.7**. But Mercury was too close to the Sun for high accuracy. Each of Venus's tracks across the Sun at different positions (due to different observation points on Earth) would take a different time to cross, and occur at different local times. So it was eventually realized that for the 1761 observations the important thing was to record the local time of the entry and exit of Venus as it passed across the Sun's disk. This was only possible if the Sun lay above the horizon for both.

The response to Halley's proposal was one of the first truly international scientific collaborations, in which at least 120 astronomers around the world agreed to attempt observations of the 1761 transit of Venus at sixty-two different places on Earth. The project was initiated by the French, in 1760, when Joseph-Nicolas Delisle, the astronomer for the French Navy, proposed Halley's scheme to the French Academy. Delisle was fiercely energetic, had met Halley before his death in London, and was connected with all the leading astronomers of Europe. He had the support of the King, who was devoted to astronomy. Delisle presented his own calculations for the durations of the transits, which were then expected in only a year's time, which differed from Halley's. He also showed the best locations on Earth to observe the transit, from Tobolsk in Siberia to the Cape in South Africa. These would be restricted to those few places on Earth whose latitude and longitude were accurately known at that time, and where the full transit occurred during daytime. The transit can last up to six hours. He sent his world map, showing the best viewing areas and transit durations to about 200 astronomers worldwide, and had it published in newspapers, becoming the center of communications for the project. Delisle devised the new, simpler method of observation, which required only the time at which the transit started or ended to be noted at two places of accurately known latitude and longitude. So essential equipment for the expeditions was an accurate pendulum clock and a large brass refracting telescope, fitted with smoked glass to allow direct observation of the Sun without blinding the astronomer. French astronomers were dispatched to Pondicherry in India, and Tobolsk in Russia.

Not to be outdone, the Royal Society in London studied Delisle's plans, and decided on full support. They chose Saint Helena (about three months' sailing time from London) and Bencoolen in Sumatra in the East Indies as sites, with less than a year to organize the expeditions and raise funding. The greatest risk was a cloudy sky at the time of the transit, and, as it turned out, the ongoing war between England and France. To deal with this, "safe passage" documents were sought from both governments. Arguing that national prestige was at stake for a project which had originated in England rather than France, the Society was successful in obtaining funding from King George II. It was understood that the improvements in navigation which would result from these observations would be good for commerce and the military. The absence of any accurate method of longitude determination at that time meant that captains tended to sail along lines of constant latitude, making ships easy to find by enemies or pirates, and, as we have noted, losses due to shipwrecks through ignorance of longitude were a very serious problem.

Neville Maskelyne was sent to the remote island of Saint Helena in the midst of the South Atlantic, where he arrived on April 5, 1761 to set up his observatory. Maskelyne was convinced of the value of lunar sightings in solving the longitude problem, and later strenuously opposed the award of the Longitude prize to the clockmaker Harrison, successfully delaying it for many years until the very end of Harrison's life. Charles Mason and Jeremiah Dixon, later responsible for surveying and mapping the Mason–Dixon line in the United States, were dispatched to Bencoolen. Four days out, they were attacked by a French frigate, and with eleven of their sailors killed and the ship badly damaged, they were forced to return to England, and this attempt was abandoned. In a second attempt they later sailed for Capetown, where they were finally successful in timing the Venus exit from the Sun soon after dawn. The entry had occurred before sunrise. There were many other observers in England, Germany, France, The Netherlands, and Scandinavia, including the Harvard Professor John Winthrop in Newfoundland, who would all see at least part of the transit, mostly viewing through telescopes and smoked glass, or projecting an image onto a wall, as Horrocks had done. As the great day approached, publicity increased, books were written, lectures on astronomy provided for the general public, telescopes sold, and the Edinburgh Magazine of April 1761 provided a detailed review of the project. Many amateurs took part, as the event became a public spectacle, and eventually a huge mass of observations were mailed to the Royal Society and French Academy for analysis following the transit. Publications in learned journals soon followed. But these showed large discrepancies between values for the solar parallax between different groups, leading to acrimonious disputes and many published corrections, which discredited the work. A general consensus developed that a much better plan could be developed for the next transit in eight years' time, based on lessons learned.

Overall, the effort had largely failed due to bad weather and many misadventures, including the effects of the war, hostile natives, and the *black drop effect* which obscures the onset of the eclipse, leading to errors of up to twenty seconds. Some longitudes were reckoned with respect to Paris, and others to Greenwich, and the difference was not accurately known. Most unlucky, for example, was the Frenchman Le Gentil, who sailed to India fully equipped, arriving first at Mauritius, where he contracted dysentery. Since Pondicherry was by then under siege by the British, he changed plans and decided to observe the transit from Rodrigues instead. Unfortunately, Rodrigues was exactly the site chosen by the French academy for another of their astronomers, Alexandra-Gui Pingre. In a comedy of errors, the captain of Pingre's ship was required to provide assistance to a French supply ship attacked by the British, he refused to stop at Rodrigues and headed for Mauritius, meaning that Pingre would miss the transit by the time he got back to Rodrigues. Meanwhile, Le Gentil had found alternative passage to Pondicherry from Mauritius on a French ship sent to relieve the Pondicherry siege, but that ship became becalmed, and with only two weeks left before the transit, he too discovered at Mahe on the Indian coast that Pondicherry was now fully under British control, so they could not dock. He returned to Mauritius, "to my great vexation," able to make only highly inaccurate observations of the transit from the rolling ship.

Le Gentil decided to wait in India for eight years for the next transit where, after further travel and misadventures, he built an observatory. When the day finally came, clouds covered the Sun, and he got nothing. He later wrote:

> I was more than two weeks in a singular dejection and almost did not have the courage to take up my pen to continue my journal; and several times it fell from my hands, when the moment came to report to France the fate of my operations.

On returning to France in 1771, he found that he had been assumed dead. His chair at the academy had been given to someone else, and his family had looted his estate. Eventually he recovered his estate and his position in the Academy, and wrote a successful book about his adventures, together with scientific observations.

For the 1769 transit, the Royal Society in London determined to try again, with the support of funding from King George II, using observations from Norway, Canada, and the Pacific. Later, it emerged that many others took part. Conditions for this second transit collaboration were greatly improved as a result of the publicity attached to the first

1761 transit observations (which educated everyone), the further development of Enlightenment ideals throughout Europe and its support for science, the enthusiasm of European, British, and Russian royalty for science, and the end of the Seven-Years War. And the trajectories of Venus across the Sun were better the second time (their last chance!), with a wider separation between tracks. Finally, telescopes had improved, with the development of the achromatic reflection telescope, which was much smaller and more portable. In France, the astronomer Jerome Lalande took the lead, publishing a new global map of favorable sites and eventually dispatching a joint French–Spanish team led by Jean-Baptiste Chappe to Baja California, just south of Arizona. The best viewing was in the South Pacific, eastern Asia, and Russia, with not much to see in Europe except in the far north. The difference in the duration of transit between observations in Lapland and in the Pacific would be about twenty-four minutes, much longer than for any of the 1761 observations. Now with strong support from royalty, national pride was at stake, and the Spanish and Russians mounted independent efforts using their own astronomers. Determined that Russia should join Enlightenment Europe, Catherine the Great managed many aspects of the eight Russian expeditions, which eventually included some European astronomers, giving detailed instructions and providing questions to the Russian Academy. Catherine then saw off the seven astronomers with a grand ceremony in the Winter Palace.

Benjamin Franklin had founded the American Physical Society (APS) in 1743. Members met in Philadelphia late in 1768 (prior to the American War of Independence with Britain in 1775) to consider their contribution to this international effort, knowing that success would improve the British colony's standing in the eyes of European scientists. David Rittenhouse led the effort, together with Franklin, who was living in London. Franklin was a Fellow of the Royal Society, and had been fully involved in London in the arrangements for the first transit expedition, and now in the second. This involved meetings with James Cook, and discussions with Maskelyne over instrumentation. The APS had great difficulty raising funds, but was finally able to support observations from Rittenhouse's farm in Norriton (where he had built two telescopes) and other local sites. Rittenhouse fainted with excitement just before the Venus entry, and so missed it, but obtained a good recording of the exit. Winthrop at Harvard was unable to raise funding, but again the impending event created much publicity, and there were many amateur observations along the eastern coast of North America. Many of these observations, including those of Dr Benjamin West in Providence, Rhode Island, were published in the first volume of *APS Transactions* in 1771, leading to an estimate of the Earth–Sun distance within 3% of the correct value.

Chappe and the French–Spanish team had to cross most of Mexico to get to Baja California, and finally had to take their precious instruments ashore in small dinghies in a storm. After succeeding with his observations, he was to die of typhus with all but a few of his expedition, two months later at the Mission San Jose del Cabo. The survivors finally made it back to Paris some months later with his notebook.

In England, Maskelyne had become Astronomer Royal, putting him in charge of both the awarding of the Longitude prize and transit expedition planning, through the Transit Committee. He provided very detailed instructions to the astronomers, from clock repair to telescope alignment and glass smoking. George III provided funds, and pre-fabricated observatories were built and tested in England (**Figure 3.6**). The astronomer William Wales was to travel to Hudson's bay, while James Cook was selected, accompanied by Charles Green and Joseph Banks, for the Pacific voyage, a mammoth undertaking full of risk. Thus began Cook's famous voyage of discovery to Australia and around the world in the *Endeavour*, with ninety-four men and 8000 pounds of sauerkraut against scurvy, which killed more sailors than the enemy at that time. Cook was given written instructions to respect any native peoples, since *"no European nation has any right to occupy any part of their country,"* and to explore the *"unknown land of the South, Terra Australis Incognita."* Cook chose the island of Tahiti (at Point Venus, the name it has retained) on which to set up his telescopes and clocks (**Figure 3.6**), one of the few places in the Pacific Ocean whose latitude and longitude were accurately known.

As Cook wrote in his diary:

This day prov'd as favourable to our purpose as we could wish, not a Clowd was to be seen the whole day and the Air was perfectly clear, so that we had every advantage we could desire in Observing the whole of the passage of the Planet Venus over the Suns disk: we very distinctly saw an Atmosphere or dusky shade round the body of the Planet which very much disturbed the times of the Contacts particularly the two internal ones. Dr. Solander observed as well as Mr. Green and my self, and we differ'd from one another in observing the times of the Contacts much more than could be expected. Mr Greens Telescope and mine were of the same Magnifying power but that of Dr Solander was greater than ours.

Figure 3.6 Portable observatory used by Captain Cook in Tahiti, showing pendulum clock. (Science Photo Library).

After many adventures, Cook returned to England and a hero's welcome, with his observations intact and a vast array of new botanical species collected by Banks. The First Fleet colonizing Australia arrived from Britain soon after, at Botany Bay in 1787.

On the day of the transit, about 250 astronomers around the world were in position at about 130 different locations, with telescopes. All the observers had difficulty detecting the onset and ending of the transit (separated by several hours) due to the "black drop" effect, which blurred the images, leading to inconsistency in their drawings. By observing the recent 2012 transit from a space-based telescope, we now know that the black drop effect is due to a combination of the aberrations of the telescope and the "limb darkening" effect, whereby the Sun is darker at its edge than in the center. Because of this, after their first examination of Cook's records in London, the Royal Society considered his mission to be a failure, and rebuked Cook severely.

The results of all the 1769 observations were eventually published by the British astronomer Thomas Hornsby in the December 1771 edition of the journal *Philosophical Transactions*. Hornsby concluded that "the mean distance from the Earth to the Sun is 93,726,900 English miles." Since the modern value is 92,955,000 miles (149,597,000 km), this error of less than 1% is truly remarkable, given the absence of photographic recordings of the transit, which were all hand drawn, and errors in longitude determination and timing due to the black drop effect. Local noon could be determined by sextant measurement of the Sun's highest altitude, to calibrate the pendulum clocks. There had been many difficulties—for example, the measurements made by several different astronomers at Greenwich of the same transit event differed by up to 53 seconds! But the increased accuracy which can be achieved by combining many measurements was not fully understood at the time. **Figure 3.7** shows a contemporary summary. You can read a full account of all the transit expeditions in Andrea Wulf's fascinating book, and a superb account of Cook and Bank's adventures in the South Pacific in Richard Holmes's beautifully told narrative, describing their life among

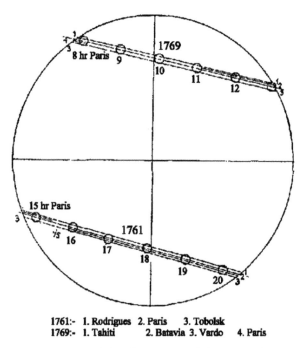

1761:- 1. Rodrigues 2. Paris   3. Tobolsk
1769:- 1. Tahiti   2. Batavia 3. Vardo   4. Paris

Figure 3.7 A summary of the observations of the 1761 and 1769 transit of Venus observations across the Sun's disk, as seen from different locations and times on Earth. The angle θ in Figure 3.5, the Venus parallax, is, for example, the angle between lines 1 and 3 measured down the page, based on the known angular diameter of the Sun. The uppermost line across from the 1769 observations was observed from Tahiti, a lower one from Paris. (From European Southern Observatory web page, transit of Venus.)

the native populations. In addition to the scientific value of the expeditions, their importance also lay in demonstrating the value and feasibility of large international collaborations, and in improving communication among nations. The project greatly increased respect for Scandinavian, Russian, and Colonial American science among Europeans. When later President of the Royal Society, Joseph Banks became a great advocate for international collaboration, since, as he put it, "The science of two nations may be at peace, while their politics are at war."

Transits also occurred in 1874 and 1882, which resulted in expeditions mounted by the British, Germans, and the Americans. The results, collated and analyzed by Simon Newcomb (Albert Michelson's supporter), further improved accuracy. Most recently, the transits of 2004 and 2012 have produced a value for the Sun–Earth distance within 0.007% of the accepted value. Perhaps more importantly, the search for new planets around distant stars (exoplanets) has benefited from these transits, since that search is often based on measurements of the reduction in intensity of light from a star due to it being blocked by a planet moving across in front of the star, exactly what happens in a Venus transit under much better understood conditions. Venus darkens our Sun to 99.9% of its unobstructed value, showing how difficult is the search for exoplanets.

The modern method of measuring interplanetary distances is based on radar echoes from the planet Venus. Our solar system consists of eight planets which formed 4.8 billion years ago from the gravitational collapse of a large molecular cloud. The inner four planets, in order from the Sun, are Mercury, Venus, Earth, and Mars, and consist of rock and metals. The next two, Jupiter and Saturn, are the largest, and are much more massive gas giants, consisting mainly of hydrogen and helium. The outermost planets, Uranus and Neptune, consist mainly of water, methane, and ammonia ice. The planetary orbits are nearly circular, and all lie in the same plane, known as the ecliptic. The Earth–Sun distance is one astronomical unit (AU) or 93,000,000 miles. The distances of the planets from the Sun (in astronomical units), and their diameters (in km) are: Mercury (0.4, 4879), Venus (0.7, 12,104), Earth (1, 12,756), Mars (1.5, 6787), Jupiter (5.2, 142,800), Saturn (9.5, 120,660), Uranus (19.2, 51,118), and Neptune (30.1, 49,528). So the largest planet, Jupiter, is more than ten times the size of the Earth, and the most distant, Neptune, is more than thirty times as far from the Sun as our Earth.

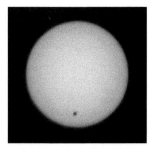

Figure 3.8 Amateur astronomer's photograph of the 2012 transit of Venus. Taken in Melbourne Australia using a 300 mm lens on June 6, 2012, between 9.45 am and 10.45 am. Nikon D7000 camera with adjustable neutral density filter to attenuate the Sun's light, fast shutter speed, small aperture. ISO 100. (Courtesy of Ashley Dunn.)

In **Figure 3.3(a)**, imagine that the Moon was replaced by Venus, so that a line from Earth to Venus and continuing on to the Sun makes a right angle at Venus. Then, if we could measure the distance from Earth to Venus, we could use Pythagoras's theorem and trigonometry to find both the distance from the Sun to Venus, and from the Sun to Earth. Then Kepler's laws, knowing the period of orbit of all of the other planets, would give us all of their distances from the Sun. The right angle occurs when the angle ϕ as seen from Earth, between the Sun and Venus, is a maximum. Then a line from Earth to Venus in the figure runs tangentially across the orbit of Venus, and this angle can be measured. When used with the measured Earth–Venus distance, this will give, by simple trigonometry, the Earth–Sun and Earth–Venus distances. In 1961, for the first time, a powerful series of radar pulses were directed toward Venus when this angle was a maximum, and a series of weak return pulses were later detected, having bounced off Venus. Knowing that they travel at the speed of light, the distance to Venus could then be accurately estimated, and from it all of the solar distances determined.

4

James Bradley, Sailing on the Thames

James Bradley (1693–1762) was born in Sherbourn, UK, and studied theology at Balliol College Oxford. His work is crucial to our story because it provided irrefutable evidence for a finite speed for light, while also producing a measurement of the time for light to travel from the Sun to Earth, within 2% of the modern value, by an entirely new astronomical method. The story is told that he had his crucial idea while sailing on the Thames, comparing the direction of the wind with that of a weather vane on his boat when it turned. The vane, which one would think would always be lined up with the constant wind direction, regardless of the direction the boat was headed, seemed to turn with the boat, even when the wind direction was steady, which was hard to understand. (This is a relative velocity problem, which we will need to explain in detail in order to understand Einstein's theory later.) His work gave much greater confidence and credibility to Roemer's earlier result, at a time when many still believed that light travelled instantaneously, or did not accept the Copernican idea that the Earth orbits the Sun.

Bradley and his wife Jane Pound had a daughter in 1745, and it was his beloved uncle, James Pound, an excellent astronomer and the rector at Wanstead, Essex, who supported him financially and fostered his interest in astronomy. Together they made many important observations for both Halley and Newton, including the positions of stars, observations of eclipses of Jupiter's moons, parallax of Mars, calculations of the orbits of comets, and measurement of the diameter of Venus, using a 200 foot telescope. When he was appointed Savilian Professor of Astronomy at Oxford in 1721 (recommended by Newton), he resigned his vicarage position, but could not afford to live in Oxford, so moved in with Pound at Wanstead, travelling up to Oxford only to give his lectures. On the death of Pound in 1724, he commenced a collaboration with Sam Molyneux, a wealthy amateur living beside Kew Gardens, in London.

Lightspeed: The ghostly Aether and the race to measure the speed of light. John C. H. Spence.
© John C. H. Spence 2020. Published in 2020 by Oxford University Press.
DOI: 10.1093/oso/9780198841968.001.0001

We are fortunate in the case of Bradley to have excellent records of exactly what he did. A century later, S.P. Rigaud, Bradley's successor as Savilian Professor, published Bradley's data. Bradley and Pound had set out to measure stellar parallax (the angle ϕ in **Figure** 3.5), the outstanding challenge of the day, in support of the Copernican idea of a Sun-centered solar system, since no parallax would be seen in a geocentric universe because the baseline is zero. Critics took the absence of any measurement of parallax to argue against the Copernican model. In this effort they failed—that angle was too small to be seen by the instruments of Bradley's day, and stellar parallax (of 0.3") would not be detected until the work of Bessel in 1838. However they did stumble on an equally important discovery. They decided to observe the nearby star Gamma Draconis (for which the parallax is less than 1"), which lay directly over the house, near the Pole star, thereby minimizing displacement of rays by refraction in the Earth's atmosphere, and which could be seen during both day and night. The Pole star is always very close to the direction of true north, the rotation axis of the Earth, not the nearby magnetic north shown by a compass. They were using a better telescope than Robert Hooke had used fifty years earlier, when he claimed to detect the parallax of the same star, however Bradley had shown that Hooke's result was erroneous.

Figure 4.1 shows their high-precision 24 foot telescope, made by George Graham (1675–1751), a famous instrument maker, which was fixed to the internal side of the chimney in Molyneux's house, extending to a hole in the roof. The angular resolution was about 1". It was found that body heat was enough to disturb the plumb line, and the motion of the plumb bob was damped by immersion in water. Air currents were reduced by enclosing the plumb line in a tube, but this provided a home for spiders and their webs which had to be removed from time to time. To make the observations, starting on December 17, 1725, James or Molyneux (who died in 1728, before Bradley had explained their observations) would lie on the couch by the fireplace looking up into the eyepiece of the telescope. A calibrated micrometer was used to swing the telescope between the star and the plumb line, and so measure the north–south deviation of the star when it crossed an east–west cross-hair. This was repeated over a period of two years, as the Earth orbited the Sun at 67,000 miles per hour (to be compared with the 186,324 miles per second speed of light and the approximately 1000 miles per hour rotational speed of the Earth at the equator), providing a total

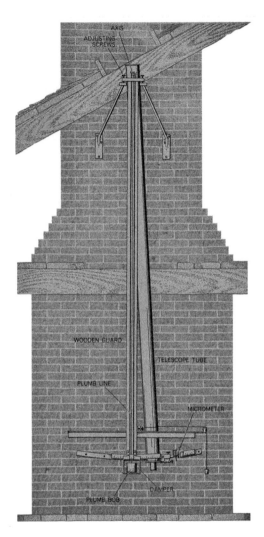

Figure 4.1 The chimney in Molyneux's house in Kew Gardens in 1725, showing Bradley's telescope attached, plumb bob, and fine motions for measuring small angles during the motion of a star. (From *Scientific American*, article by A.B.Stewart (1964)).

of eighty observations. Notice that they were only measuring the angular northerly *change* in the position of the star, using the local vertical (plumb line) as a reference. These changes were as large as 40".

The instrument proved remarkably rugged—Molyneux wrote in his notebook on one occasion:

> *This was a rainy, blowing tempestuous night, however this morning trying the instrument again we found the index as we had left it at 5[seconds of arc], and we were obliged to alter it only to 6, so that with this blowing, sudden change of weather, it alters but one second to bring it to the plumb line.*

Eventually, Bradley realized that these stellar motions could not be due to parallax because the displacement observed was in the opposite direction to that expected by parallax. (We will see that parallax depends on the distance to the star, whereas the aberration effect which Bradley discovered does not.) By comparing Gamma Draconis's motion with that of another star, he was also able to exclude nutation (or wobble) of the Earth's rotation axis. They also considered and excluded atmospheric refraction as a cause by comparing results from different stars.

In 1727 Bradley had Graham build him a new more accurate telescope, to be installed at his aunt's house at Wanstead, and observations began in August. This was half the length of the previous one but could be adjusted more rapidly, with a smallest angular increment of 0.5" and a much wider field of view (more than 6° around their zenith) to allow him to test his ideas on the dozen stars which could then be seen throughout the year and during both day and night. He discovered a general pattern to the motion of the stars:

> *The apparent motion of every one tended the same way ... they all moved southward, while they passed in the day, and northward in the night, so that each was farthest north when it came six of the clock in the evening, and farthest south when it came about six in the morning.*

The deviations varied sinusoidally throughout the year. Rigaud writes that Bradley's insight based on his sailing trip to explain the effect occurred in the fall of 1728, and is described by Thomas Thomson in his 1812 history of the Royal Society:

> *At last, when he despaired of being able to account for his observations, a satisfactory explanation occurred to him all at once ... he accompanied a pleasure party on a sail upon the river Thames. The boat had a mast, to which a vane was attached at the top. Dr Bradley remarked, that every time the boat put about, the vane shifted a little, as if there had been slight change in the direction of the wind. The sailors told him that the wind had not shifted, but that the apparent change was owing to the change in the direction of the boat ... this accidental observation led him to conclude that the phenomenon which had puzzled him so much was owing to the combined motion of light and of the earth.*

Evidently the same effect was occurring in Bradley's stellar observations, with the light playing the role of the wind, and the Earth moving instead of the boat. As a dedicated fast trapeze dinghy sailor until recently, like all yachtsmen, I know that the faster you go, the more the apparent wind direction appears to swing around and come increasingly from the forward direction, an effect which is particularly evident on fast planing dinghies, wind surfers, and ice boats. Similarly, whereas rain appears to fall vertically on a windless day, it becomes slanted as we walk rapidly forward, requiring an umbrella to be tilted forward.

Figure 4.2(a) shows a telescope on a stationary Earth (with respect to the Sun) pointed directly at a star. The motion of the star relative to the Earth does not affect the angle at which the starlight reaches the Earth because the star is so far away. We can think of the photons from the star as bullets running down the center of the telescope, its optic axis. Now we must allow the Earth to move sideways, which it does in solar orbit at about 67,000 mph (30 km/s), at times directly across the direction of the incoming starlight. With the vertically mounted telescope, the telescope will move sideways during the time the bullets are running down the axis of the telescope, and so hit the sides of the telescope, and not be seen by the astronomer. What we need to do in order to keep the bullets on the axis (and so see the star) is to tilt the telescope as shown in **Figure 4.**2(b). Then as it moves to the left due to the Earth's motion around the Sun, which is much faster than its rotational speed, the bullets or photons of starlight will be kept running along the axis and the star will be seen by the viewer. Because the telescope is tilted through a small angle, it will appear to lie at the apparent position shown, in the direction of the tilted telescope, not its position if it were observed from the stationary Earth in **Figure 4.**2(a). If the Earth's speed (about 67,000 mph around the Sun) is v, then during the time Δt it takes the photons to run down the length $c\Delta t$ of the telescope, it will have moved sideways by $v\Delta t$. The angle of tilt is $\tan \theta = v\Delta t/c\Delta t = v/c$. This is the aberration of starlight effect which Bradley saw and correctly analyzed non-relativistically. The importance of it is that from a measurement of this angle, we either use our knowledge of the speed of the Earth's orbit around the Sun to find the speed of light, or vice versa.

Finally we must consider how this angle will vary throughout the year. It may be helpful to imagine that we are viewing a star in the northern hemisphere near the Pole star, as Bradley did, which always lies in the direction of the rotation axis of the Earth, and always in the

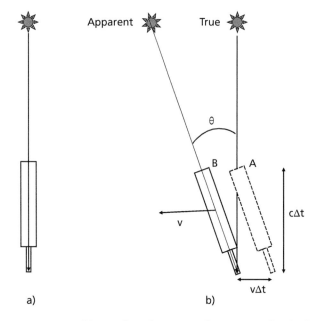

Apparent   True

θ

B   A

cΔt

v

vΔt

a)   b)

Figure 4.2 Because of the Earth's velocity v, a telescope must be tilted to keep the starlight coming down the axis and not hitting the sides of the telescope. (a) shows a telescope on a stationary Earth and (b) on a moving Earth.

direction of due north. Then it can be shown that the telescope in **Figure 4.2**(b) traces out a cone throughout the year around the true direction of the star, as Bradley discovered. After six months, the Earth has reversed the direction of its velocity with respect to the Sun.

This is also an example of a general *relative velocity* problem, which we will need to discuss in much more detail later, so we may as well establish the ideas now. These problems are familiar to all science undergraduates, usually taught in the form of a swimmer trying to cross a flowing river. The exam question is usually "find the direction in which the swimmer must head in order to cross the river in the shortest time." It's an important problem for study of the speed of light, since it will also arise in our chapter on Albert Michelson and his famous interferometer, and in Einstein's theory of relativity.

Pilots, viewing the track of another aircraft, are also familiar with the deceptive nature of relative velocity issues, as I've learnt flying large gliders in the plentiful thermals over the Arizona desert, where it can be hard to make sense of the track of aircraft flying across and below you. **Figure 4.3** shows light again from a star falling from the true star position onto the Earth. Like the dinghy, the Earth is moving to the left, so the direction of the light, like the wind, swings around to appear to come from the apparent star position. The angle θ is the stellar aberration angle measured by Bradley.

More familiar than the apparent wind direction that sailors deal with are the moving walkways at airports. If you walk along the belt at 3 mph, and the belt is going at 4 mph, your speed relative to the ground is obviously 7 mph. By subtraction, your speed relative to the belt is equal to your speed relative to the ground (7 mph) minus the speed of the belt relative to the ground (4 mph). In the theory of relativity, this relationship is given the fancy name of a Galilean transformation, and the belt is called an *inertial frame*, because it moves at constant speed relative to the ground.

This simple one-dimensional result carries over into three-dimensional motion, where we must use vectors, but the idea is the same, so that, for the aberration of starlight, if we take the Sun to be equivalent to the ground (a "stationary" point), then the belt is the Earth moving with velocity $\mathbf{V}_E$ relative to the Sun, the passenger is the light with velocity $\mathbf{V}_L$ relative to the Sun, while $\mathbf{V}_{LE}$ is the velocity of the light with respect to the Earth, which we want to find. Using vectors we have the same subtraction as for the walkway for the triangle shown in **Figure 4.3**, where

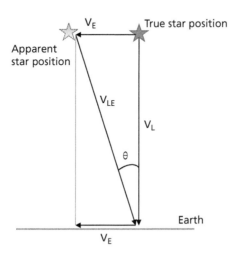

Figure 4.3 Relative velocity vectors for light from a star.

$$V_{LE} = V_L - V_E \qquad .$$

The light, like the wind approaching a speeding yacht on a broad reach, is seen to appear to come from a position ahead of the star, at the angle θ, as measured by Bradley, where

$$\tan\theta = |V_E| \,/\, |V_L| = v\,/\,c = \beta$$

as we obtained previously. Here the velocity of light is represented by the symbol c (an international convention), the Earth's speed by v, and we have introduced another internationally agreed upon symbol β, which will become important in relativity theory. In 1905, Einstein added a further small correction to this equation.

This mathematical interpretation of Bradley's data fitted well with all his observations from stars, taking into account the changes in the Earth's orbital velocity with the seasons. His best estimate of the average value of θ (when the Earth is moving at right angles to a line from the Earth to the star, as in **Figure 4.3**) was 20.25 seconds of arc, compared with today's value of 20.47", a remarkable achievement.

Bradley's equation gives the ratio of the speed of light to that of the Earth, and he calculated β to be about 1/10210. Using the best estimate of the Earth–Sun distance available, he was then able to estimate the time light takes to travel across the diameter of the Earth's orbit around the Sun, and found it to be about sixteen minutes and twenty-six seconds, not so different from Roemer's estimate of twenty-two minutes, and very close to today's value of sixteen minutes and forty seconds.

Bradley's achievement was later described by Sir Arthur Eddington: "It was only by extraordinary perseverance and perspicuity that Bradley was able to explain the phenomenon." As A.B. Stewart points out, his work was important in three ways. First, it demonstrated the value of quantitative analysis and its use in discrediting the geocentric view of the universe. Second, it established a new standard of precision in astronomical instrumentation. Finally, it gave confidence to the idea that light travels at a finite velocity and provided the groundwork for the theory of relativity. Today, the measurement of the aberration of starlight is one of the simplest ways to measure the velocity of the Earth around the Sun.

Bradley's work was improved upon in subsequent years, particularly by F. Struve, who in 1843 published a very accurate value for the aberration of starlight (20.4451 seconds of arc), from which a time of 497.8 $+/-$ 3 seconds could be deduced for the time of sunlight to reach the Earth.

We will see in later chapters that this work became relevant to the debate over the existence of the Aether. In the early nineteenth century, Augustin Fresnel had shown that a wave theory of light supporting Bradley's results was consistent with the existence of an Aether medium to support lightwaves, filling all space, and through which the Earth moved. Michelson and Morley, whose story we will take up in a later chapter, showed in 1887 that if there is an Aether, it must move with the Earth. Only Einstein's theory of relativity in 1905 would resolve this dilemma completely. Bradley's work, which was intended to support the Copernican universe with its stationary Sun, ended up being used to support Einstein's theory in which all motion is relative. Bradley's measurement was nevertheless the first observational support for Copernicus's theory.

5

The Nineteenth Century
Light Beams Across the Rooftops of Paris

Until now, we have explored the early attempts to measure the speed of light, but we have barely touched on an important question—the nature of light itself. The ancient Greeks made great progress using simple geometric arguments, in which light travelled in straight lines as particles travel, but by the late seventeenth century there had begun to accumulate more and more evidence that light might be a kind of wave. We discuss this wave–particle paradox in more detail in later chapters, however to preserve our broadly chronological tale, we need to make a brief diversion into the history of the discovery of the wave nature of light, due to Huygens, Young, and Fresnel.

In 1690, Huygens had been led to his wave theory of light and a model of the Aether which supports it by noting how, if a line of hard balls in contact with each other is struck at one end (as shown by the "Newton's cradle" toys sold in Science Museum stores), the pulse of movement emerges at the other end apparently almost instantly. With Roemer's results in mind, he therefore assumed an Aether, filling all space with minute, closely packed hard elastic spheres in contact like tiny invisible ball bearings, through which a disturbance could propagate in any direction. With the experience of the wavefronts which appear when a stone is dropped into a still pond in mind, he realized that by taking any point on a wavefront as a new source of a spherical wave, he could explain Snell's law for refraction and other optical phenomena, such as birefringence. This occurs when a material appears to have two different simultaneous values of refractive index, the effect discovered by Roemer's mentor Bartholin.

These ideas, and new interference effects observed by Thomas Young in 1801, were taken up and given mathematical form by Augustin Fresnel (1788–1827), with such great explanatory power over such a wide variety of optical effects, that by the end of the nineteenth century,

Lightspeed: The ghostly Aether and the race to measure the speed of light. John C. H. Spence.
© John C. H. Spence 2020. Published in 2020 by Oxford University Press.
DOI: 10.1093/oso/9780198841968.001.0001

Newton's corpuscular model for light as a series of tiny bullets had been almost completely replaced by the wave theory.

In his analysis of the interference rings between two glass plates ("Newton's rings," discussed later), and of colors in soap bubbles, Newton had, however, in 1672, hinted at the possibility of a periodic nature for light, in correspondence with Robert Hooke, an ardent supporter of the wave model. Hooke, prior to Roemer's publications, had written that he knew of no experiments to indicate that light travels instantaneously, and seemed to favor a very large finite velocity. After Roemer's paper, he was unconvinced of Roemer's result and wrote in support of an infinite velocity. Later, in 1801, Thomas Young (1773–1829) was to explain his interference fringes using a wave model, and provide the first convincing experimental evidence that light was a wave, by estimating its wavelength.

Young was a Quaker and a remarkable polymath who spoke Greek and Latin at the age of fourteen, and later many other languages, becoming a Fellow of the Royal Society at the age of twenty-one. He started out as a physician and soon after became independently wealthy by inheritance. He was appointed Professor of Natural Philosophy at the Royal Institution in 1801, Foreign Secretary of the Royal Society in 1802, and a Foreign Member of both the American Academy of Arts and Science and of the French Academy. Herschel considered him to be a genius, as he appeared to be able to answer any kind of question in any field of science. Apart from his observation of interference between light waves, he is remembered also for his modulus of elasticity, his work on the basis of color vision, and his correct explanation for the accommodation of the human eye, how it can re-focus on things at different distances from us. He also explained surface tension and contact angles, and worked on the definition of energy (defining kinetic energy correctly for the first time), the theory of tidal motion, and a method of tuning musical instruments ("Young temperament"). In fierce competition with the French expert Champollion, Young achieved fame by deciphering Egyptian hieroglyphics, and in 1814 he translated the Rosetta Stone, as described in his Encyclopedia Britannica article of 1818. But Young was unlucky in having to compete with the dynamic Humphry Davy in giving his Royal Institution lectures—Davy was an excellent lecturer who drew large crowds, unlike Young, whose style was dry and understated. But his enduring fame results from his demonstration of interference between two beams of light, his clear statement of the principle of superposition, and his estimate of the wavelength of light.

Fuzzy shadow edges cast by perfectly straight edges such as a sword had puzzled scientists for a long time. How to explain this, if light consisted of particles which could only travel in straight lines? On the other hand, if light really was a wave, similar to water and sound waves, why did it not go around corners like them?

In around 1650, Grimaldi, a Jesuit priest, had described these interference fringes (parallel bands of bright and dark light) at the edge of the shadow of a blade, prior to Newton's similar unexplained observations. Young became convinced that these fuzzy shadow edges were interference effects, and hence evidence for the wave nature of light. To make these effects strong, the key requirement is to have a very small source of light—which becomes weak—or a big one like the Sun a very long way away.

Young commenced his research with acoustics and water waves, which provided him with the analogy he needed to support Huygens's 1678 wave theory of light. This treated light as a series of pulses rather than a continuous wave, but used spherical wavefronts. Young was the first to understand the principle of "linear superposition," meaning that water waves which cross, coming from different directions, will add in height, and that this idea can be used to explain tidal flows, but also applied to sound and light waves. All of this he described in his paper, read to the Royal Society in November 1801, which also explained the colors and interference effects seen in Newton's rings and thin soap films in terms of the frequency and wavelength of light. Newton's rings are circles of interference fringes of different colors seen when a plano-convex lens is placed on a reflecting surface, irradiated by sunlight. Because this geometry is very un-demanding on the coherence properties of the light, and could be obtained with the lenses of the day, it was the first really strong evidence of interference phenomena.

Newton, who explained tides as due to the gravitational forces of the Moon and Sun, had used a limited form of this idea to explain the tides at Batsha bay in Vietnam, where travellers reported the strange phenomenon of a completely static water level for one entire day every fourteen days, between which there was only a single slow tide, increasing and falling. Normally, due to the gravitational effects of the Moon and Sun, there are two high tides and two low tides each day. In his *Principia* (1688), Newton explains the Batsha tide as the summation of two different tidal flows (one from the China Sea, and one from the Indian Ocean) arriving at different times, which could cancel out twice in each lunar month. This is the underlying principle of interference, and Young was the first to spell it out clearly for light waves.

These early researchers (including Newton with his rings) used sunlight as a source of light, which contains many colors or frequencies, so that an understanding of interference effects at this time was much confused by the contributions from light of different colors. In general, waves of different frequency cannot interfere, except very briefly, when they "beat." We hear these beats when a piano is being tuned. Each note in the middle register of a piano consists of three strings playing the same note (to make it louder), all struck by the same hammer. (Two strings are used for the higher notes, one for the lowest notes.) During tune-up, if one of the three is slightly out of tune, we hear a rapid periodic modulation in the volume coming from the strings, whose frequency is equal to the difference in the resonant frequency of the two out-of-tune strings. This is known as beating. The piano tuner reduces this beat to zero when the strings are all in tune, by adjusting the tension of the strings. Similar things can be done today using two different lasers tuned to nearly the same frequency, which beat slowly, allowing the observation of interference effects within the period of one beat, if we use a camera with a very fast shutter which is open only during the beat. For a thin oil film, each color can interfere with itself in a particular direction, and Young was in 1801 able to explain the colors seen reflected by a thin oil slick using his wave theory of interference.

It is interesting to see what Newton said about the possibility of light being a wave. After all, he had produced his *Newton's rings* interference fringes, which can only be explained on a wave theory. Here is what he writes in his book Opticks:

> *If a stone be thrown into stagnating water, the waves excited thereby continue to arise in the place where the stone fell into the water, and are propagated from thence in concentric circles upon the surface of the water to great distances. And the vibrations or tremors excited by vibrations in the air by percussion continue a little time from the place of percussion in concentric spheres to great distances. And in like manner, when a ray of light falls on the surface of any pellucid* (a transparent and/or reflecting) *body and is there refracted or reflected, may not waves of vibration, or tremors, be thereby excited in the refracting or reflecting medium at the point of incidence . . . and are not these vibrations propagated from the point of incidence to great distances? And do they not overtake the rays of light, and by*

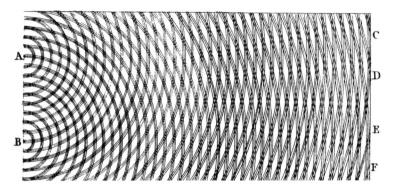

Figure 5.1 The drawing Young published to show interference between waves from two different small sources at A and B. The interference is constructive around D and E. (The sources A and B could alternatively be points where two small stones hit a still pond at the same time.) (From A. Robinson (2006), Fig 7.4.) (From Science Photo Library).

overtaking them successively do they not put them into the fits of easy reflection and easy transmission and easy reflection described above.

We can see that Newton was undecided about the issue—here he is suggesting both a particle ("*ray*") and wave model working together, with a surprising similarity to the recent "pilot wave" hidden variable theories proposed by de Broglie and Bohm in response to the "EPR" paper of Einstein, which we discuss in a later chapter, where waves guide particles.

Young moved on to the interference effects he observed between water waves excited at two points in a shallow trough, as shown in his drawing in **Figure 5.1**, which he produced using a ripple tank (as still used today for lecture demonstrations) in another lecture in 1802 at the Royal Institution. In that lecture he also discussed Newton's rings. As Andrew Robinson describes in his biography of Young, knowing the curvature of the glass used to obtain the rings, he was able to estimate correctly the wavelength of red light as 0.0000266 inches (0.65 microns), and also that of the other colors seen, a remarkable achievement. Young also wrote, in 1802, that he had discovered a general law:

Wherever two portions of the same light arrive at the eye by different routes, either exactly or very nearly in the same direction, the light becomes most intense when the difference of the routes is any multiple of a certain length, and least intense in the intermediate state of the interfering portions; and this length is different for light of different colors

The "certain length" would turn out to be the wavelength of the light, and this statement became the principle on which is based the science of interferometry, from Michelson's instrument, to modern electron and neutron beam interferometers and the LIGO interferometer used recently to detect gravitational waves. But Young's greatest achievement was to demonstrate, in his November 1803 lecture *Experiments and Calculations Relative to Physical Optics*, interference effects on a screen using sunlight passing around the two sides of a very narrow strip of card placed within a beam of light. Here is how he describes his experiment:

> *I made a small hole in a window shutter, and covered it with a piece of thick paper, which I perforated with a fine needle. For greater convenience of observation, I placed a small looking glass without the window shutter in such a position as to reflect the sun's light in a direction nearly horizontal, upon the opposite wall, and to cause the cone of diverging light to pass over the table on which were several little screens of card paper. I brought into the sunbeam a slip of card about one thirtieth of an inch in breadth, and observed its shadow, either on the other wall or on cards held at different distances. Beside the fringes of color on each side of the shadow, the shadow itself was divided by similar parallel fringes, of smaller dimensions... Now these fringes were the joint effects of the portions of the light passing one each side of the slip of card, and inflected, or rather diffracted, into the shadow. For, a little screen being placed a few inches from the card, so as to receive either edge of the shadow on its margin, all the fringes which had before been observed in the shadow on the wall disappeared...*

In other words, the interference fringes in the shadow region caused by light passing around either side of the card and overlapping at the viewing screen disappeared if he blocked the passage of light on one side of the card, demonstrating interference under controlled conditions for the first time, with the correct explanation. The light waves going around the sides of the card recombined like Newton's tides at Batsha. This experiment worked partly because he used "a *fine* needle" to perforate the opaque screen illuminated by sunlight covering his window. This would have increased the coherence of the light, improving its ability to produce interference fringes. But how did he know to do this? A very dark room would then be needed to see the interference fringes, and it would have worked better if his looking glass had focusing properties, giving more light. Young once again estimated the wavelength of light from these fringes, in agreement with his values from Newton's rings.

Sometime between 1803 and the publication of his *Course of Lectures on Natural Philosophy and the Mechanical Arts* in 1807, where he describes the work, Young probably performed the next experiment, which has earned him immortality—*Young's slit experiment*. He says he replaced the narrow strip of card used previously with a card wider than the light beam, blocking it completely, but containing two very fine pinpricks, both passing through the card, and very close together. On the far wall, in the dark, he saw a set of bright parallel stripes (interference fringes) running at right angles to a line between the holes. Again somehow he understood the importance of using the smallest possible hole in the screen covering the window of his completely dark room. As noted previously, this makes the sunlight more coherent and capable of producing interference effects, but also makes it weaker. His thirty-ninth lecture (still in print) describes this most famous experiment, in which the two pinpricks are exactly analogous to the points A and B in **Figure 5.1** for the water waves, acting as point sources of water waves or light. The alternating dark and bright stripes on his screen had a central bright stripe. For this stripe, the distance to each of the pinholes is equal, so the waves are in phase and add constructively, rather than cancelling.

If you try to see these interference fringes at a sharp shadow edge in sunlight, you'll find it very difficult to repeat Young's experiment as he did it—you would need to work in a very dark room with a tiny hole in one window onto which the sun is focused from outside. Much easier using the light from a laser pointer, which is a single color or frequency! There is some disagreement among historians of science as to whether Young actually did this experiment (unlike the first one) since, although he describes the experiment and provides a sketch of the flow of energy from the pinholes to the screen, he provides no qualitative results. Nor does he describe his light source or say explicitly that he did it. He writes, "The simplest case appears to be when a beam of homogeneous light falls on a screen in which there are two very small holes or slits..." and goes on to describe the experiment and results expected (or observed) in detail.

Most mysterious of all for this experiment, as we now know, is the fact that if, using modern apparatus, we repeat Young's pinhole experiment by sending in one photon (or electron) every minute (or every hour, or week), so they land like raindrops at the detector, we still get the same pattern of stripes. But how does each particle know where the previous one went, in order to build up the right pattern? Again, light seems to be travelling as a wave but arriving as a particle, giving a finite speed to light. For these reasons, Young's lectures of 1807 had described the experiment which Richard Feynman describes as "containing all the mystery of quantum mechanics."

But Thomas Young had enemies, mostly those under Newton's spell (such as Henry Brougham, an influential reviewer who wrote scornfully of his work). Young's laconic, dry, and tedious rebuttals did not help. And so it was Augustin-Jean Fresnel, working independently in Paris from about 1815, who was the next important figure after Young and Huygens to build acceptance for the wave theory of light. He constructed the first general wave theory of light, formulated in mathematical terms, which has stood the test of time, as we now briefly summarize. He is also important to our story for his proposal of the Aether drag hypothesis, a powerful idea throughout the nineteenth century, tested by Hippolyte Fizeau in Paris, Albert Michelson in Cleveland, Ohio, and others, leading directly to Einstein's theory. Fresnel did so in one of the most dramatic moments in the history of science, winning a great prize against an almost entirely hostile panel of referees and the immense authority of Newton, with a single decisive, irrefutable, experiment.

Fresnel was born in 1788 in Eure in Normandy into a family of successful public servants, military officers, and diplomats. Their mother's younger brother was father to the writer Prosper Merimee; she had home-schooled the four Fresnel sons, and was a firm believer in the Jansenite Catholic religion, which emphasized original sin. Augustin was also a devoted believer and, despite steadily declining health, worked intensely with a keen sense of duty to others to repay God for his intellectual gifts. After an undistinguished childhood Augustin did well at the Ecole Polytechnique and eventually joined the *Corps des Ponts* (bridge and road engineering), in which he remained for life.

In March 1815, the year Napoleon came back from Elba, Fresnel, a Royalist, joined a small, unsuccessful military force opposing Napoleon on his way to Paris. As a result he lost his government appointment and was put under police surveillance. He moved in with his mother near Caen in Normandy, and it was during this enforced leisure that he started his experiments in optics and diffraction, with no formal prior training.

Following the Battle of Waterloo (June 18, 1815) he got his old job back under Louis XVIII, and so was permitted to work in Paris. In June 1819 he was appointed by Arago to serve on the commission on

Lighthouses, where he revolutionized their design, but contracted tuberculosis in around 1824, leading to his death in 1827 at the young age of thirty-nine.

Fresnel's achievements in applying a mathematical formulation of the new wave theory, against the authority of Newton (and undertaking new experiments of his own), to explain the outstanding optics problems of the day are remarkable. These included diffraction—the blurring seen at the edge of a shadow, where fine bright and dark bands of light could be seen running parallel to the shadow edge. He provided an explanation for the polarization effects discovered by Malus, Arago, Biot, and Brewster, explained Newton's rings, and provided support for Young's interpretation of his experiments. Young's important 1801 paper was written in English, which Fresnel unfortunately could not read, but which he had translated for him by his friend Arago. In 1818 Fresnel produced the mathematical theory of near-field diffraction, with full account of phase differences, in almost exactly the form it now appears in modern textbooks.

In 1817, the French Academie des Sciences announced that diffraction would be the topic of the 1819 Grand Prix. Practically all of the committee (which included Arago), including Poisson, Biot and LaPlace were corpusculists, who, following Newton, believed that light was a stream of particles. There were only two entries, Fresnel's theory of diffraction and one other. To discredit Fresnel's theory, Poisson used it to predict that just beyond a coin, illuminated face-on by a collimated beam of coherent light, there should be a bright spot in the center—clearly absurd in this darkest part of the shadow of an opaque disk. When Arago demonstrated that this bright spot actually occurs (as still demonstrated in optics courses today at universities around the world), the committee unanimously awarded the prize to Fresnel. Like the coherent light waves travelling around the sides of the card held edge-on in Young's experiment, the bright spot is easily explained today by constructive interference between light waves travelling around the edges of the coin and beyond, propagating back toward the center. (Recent historical research, summarized by S. Ganci (2013), has shown that the bright spot had actually been reported in the scientific literature several times before, as early as 1723, in work overlooked by Poisson's committee.)

Fresnel went on to achieve greater things, including his analysis with Arago showing that the disturbance responsible for light waves was purely transverse. Two of the most important kinds of waves are known as transverse and longitudinal. For transverse waves, like ocean waves, the undulations move up and down, and this is the direction of polarization. These undulations occur at right angles to the direction in which the wave is travelling, the direction in which wave crests run. For longitudinal waves, such as sound waves, however, the molecules oscillate along the direction in which the wave is propagating. A birefringent crystal can pick out transverse waves of light whose vibrations oscillate in two directions at right angles to each other, with both at right angles to the direction of propagation. Fresnel and Arago found that no interference was produced by this orthogonally polarized light, obtainable from birefringent crystals. Young had also discussed this in the 1818 edition of *Encylopaedia Britannica*, proposing both longitudinal and transverse components. Fresnel was able to later extend his theory to circular and elliptical polarization effects, and to explain total internal reflection and double refraction. By firmly establishing that light was a transverse wave, Fresnel created a serious problem for supporters of the Aether thesis. As an invisible elastic solid, the Aether could be expected to support both longitudinal and transverse waves, like all known elastic solids, however no evidence was ever found for longitudinal light waves.

The personal relationship between Young and Fresnel was complicated. Fresnel, unaware of Young's work, had rediscovered interference effects in 1815, but acknowledged Young's priority in a letter to him in 1816. In turn, Young supported the award of the Rumford Medal to Fresnel. And Arago, Fresnel's senior, who translated Young's papers for Fresnel, became good friends with Young. Following Fresnel's death, Arago in his memoir of him writes vividly of his encounter with Young, and particularly his wife:

In the year 1816, I passed over to England with my learned friend M. Gay-Lussac. Fresnel had then just entered in the most brilliant manner into the career of science by publishing his "Memoire sur la Diffraction". This work . . . became the first object of our communication with Dr. Young. We were astonished at the numerous restrictions he put upon our commendations, and in the end he told us that the experiment about which we made so much ado was published in his own work on Natural Philosophy as early as 1807. This assertion did not appear to us correct, and this rendered the discussion long and minute. Mrs Young was present, and did not appear to take any interest in the conversation, but, as we know, that fear, however puerile, of passing for learned ladies - of being designated blue-stockings - made the English ladies very reserved in the presence of strangers, our want of politeness did not strike us till the moment Mrs Young rose up suddenly and left the room. We immediately offered our most urgent apologies to her husband, when Mrs Young returned, with an enormous quarto under her arm. It was the first volume of the Natural Philosophy. She placed it on the table, opened it without saying a word, and pointed with her finger to a figure where the curved line of the diffracted bands, on which the discussion turned, appeared theoretically established.

Arago, writing here not long after the French Revolution (*Liberte! Fraternite! Egalite!*), and a liberal Republican at heart, perhaps wants us to understand that the educated ("blue-stocking") ladies of France were more "liberated" than their English equivalents. Fresnel is best remembered, as on the monument to him at his birthplace, for his stepped lenses for lighthouses (*Fresnel lenses*), which saved many lives at sea, and for which he was most proud, as described in Levitt's 2013 book. He was elected a member of the Académie des Sciences in 1823, a Knight of the Legion of Honor in 1824, and a Foreign Member of the Royal Society in 1825. He was awarded that society's Rumford medal in 1827, all of these honors indicating the steady acceptance of the wave theory of light in both France and England. He died in 1827.

Returning now to the question of the speed of light, all this work assumed that the Aether, in which light was a disturbance, was stationary with respect to the Earth, meaning that it would have to rotate around the universe daily with the Earth, an unlikely prospect. If the Aether were fixed to distant stars, the question of what the speed of light should be in a moving medium became crucial, as for the case of a light source on Earth moving through the surrounding stationary Aether. Was there then an "Aether wind" which the Earth, and earthbound light sources moved through as the Earth rotated? If so, in which direction did it blow?

Fresnel proposed his Aether drag hypothesis to resolve this dilemma in 1818. Arago had sought a variation in the refraction angles of light from stars due to the differing stellar (and hence light) velocities, but had not found it. As we saw in Chapter 1, the velocity of light v in glass with refractive index n had been taken by Snell's law as

$$v = c \,/\, n \,.$$

As discussed in more detail in Chapter 7, Fresnel now proposed that Snell's law be altered to explain Arago's finding by adding a correction (linear in v) to the velocity of the light in the glass, caused by the entrained Aether. Fresnel's idea was that the glass carries some of the Aether along with it, leading to a higher Aether density within the glass. He took the density of the Aether in the air to be ρ_a, and in the medium, ρ_m, so that light would be dragged by a factor $v_d = v$ $(1 - \rho_a/\rho_m)$, where v is the velocity of the glass through the Aether. Assuming $\rho_a/\rho_m = 1/n^2$, his expression for the speed of light in a medium due to Aether drag became

$$V = c/n + v\left(1 - 1/n^2 \right).$$

His correction became known as *Fresnel's Aether drag coefficient* and launched attempts throughout the remainder of the nineteenth century to measure it, and so confirm or reject his theory. The experimental *confirmation* of Fresnel's correction by Fizeau later in the century greatly confused understanding, for as it happens, Fresnel's Aether drag coefficient gives the right answer for the wrong reasons. The correct derivation of the above equation had to wait for Einstein.

The first experiments to make terrestrial measurements of the speed of light were those of Fizeau and Foucault in Paris in around 1860. These were eventually highly successful, providing measurements within less than 1% of modern values. However, their work was based on an important earlier proposal by Arago using rotating mirrors, and on an overlooked experiment published by the British physicist Charles Wheatstone (1802–75) in 1834, who may have been the first to suggest the use of rotating mirrors, and whose ingenious method anticipates the modern streak camera.

Wheatstone was a prolific inventor, and arguably the inventor of the first practical telegraph. He started out in business in London as a musical instrument maker, making concertinas in around 1823; also publishing papers on acoustics. His scientific achievements, now almost completely overlooked, were remarkable and prolific, while what he is most remembered for (the *Wheatstone bridge*) was actually invented by Christy and analyzed and promoted by Wheatstone. His early work in acoustics resulted in publications of the Chladni figures, formed on a membrane or drum covered with sand when it is set vibrating. He went on to provide the theory for stereoscopic fusion of images from different directions and to give a spectral analysis of the light from sparks, and remarkably, to suggest that it could be used to identify the elements in the electrodes, as is done using similar methods today.

While the origins of the telegraph remain obscure, and depend on a judgement on the degree of sophistication on the encoding of signals (of which Morse code was the most successful), Wheatstone can certainly lay claim to being the inventor of the first practical telegraph made available for public use. This was the two-needle telegraph which he developed with Cooke. In 1840 he improved it to show alphabet letters directly in response to electrical pulses sent down a wire, which led to a printing telegraph in 1841 and finally to an automatically transmitting and receiving instrument. He then moved on to submarine telegraph research in Swansea Bay and further research into improved dynamos, coming up with, independently but at the same time, the continuous self-exciting Siemens design of 1867. He also invented a method of synchronizing clocks using electrical signals.

It was natural to ask how fast these telegraph signals were going along the wires which were soon laid across the Atlantic Ocean. As Professor of Physics, he set up half a mile of copper wire in the basement of King's College in London to measure "the speed of electricity," which we now know is the speed of light.

Wheatstone's apparatus is shown in **Figure 5.2(a)**. His aim was to measure the time it takes for an electrical pulse created by a spark to run along half a mile of wire. To do this, he compared the light from the spark as it was generated with a spark produced by the same electrical pulse after it had been sent on a half-mile detour through a long wire. The light from these two sparks was viewed though a "whirling mirror," which rotated about a vertical axis, producing a streak of light appearing to the viewer as a short bright horizontal line. The delayed streak was displaced to lie parallel to and above the immediate streak. If the streaks lined up at start and finish, the electrical signal must have travelled instantaneously.

The rotational speed of the mirror was about 800 revolutions per second, which he determined from the pitch of the sound emitted, compared to a piano playing a note one octave above the G# in the middle of the keyboard. The spark streaks were seen over an angular range of about half a degree, which he claimed gave a time resolution of better than a microsecond, good enough to resolve the time delay over half a mile. Wheatstone's experiment was remarkable, since he clearly shows a retardation in the streak from the delayed spark, and from this was able to estimate the speed of electrical transmission. He found this to be 250,000 miles per second, somewhat greater than the speed of light—186,000 miles per second! It is not clear what caused this error.

Wheatstone married Emma West in 1847, with whom he had five children. He had a "morbid timidity" in front of audiences, and was

(a)

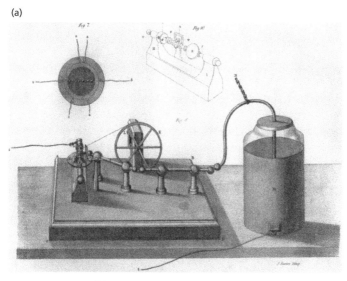

(b)

Figure 5.2 (a) Wheatstone's apparatus for measuring the speed of electricity. The Leyden jar at right is a capacitor which creates sparks viewed by a mirror which is rotated by a "whirling machine." (From C. Wheatstone, *Phil Trans* 124, 583 (1834).) (b) Wheatstone's "photometer," a clockwork rotating mirror apparatus.

rather indifferent at teaching at King's, which he avoided. His books and instruments were left to King's, where they can be seen today. I was lucky to be allowed access to the collection of his scientific instruments, one of which was the rotating mirror apparatus ("photometer") shown in **Figure 5.2(b)**. The mirror can be seen within the ring at the top. It rotates about a vertical axis.. The mirror is driven by clockwork, like a modified carriage clock without an escapement to slow the motion. The clockwork mechanism beneath the dial also indicates the total revolutions turned. The photometer came in a beautiful mahogany case. The instrument, being made of brass, looked almost new. In the corner of the case was mounted the key, which I carefully removed to wind up the mechanism. It worked perfectly after 184 years. The top dial, which advances continually during running, shows the total number of revolutions of the little mirror, so I could estimate the speed to be a few hundred revolutions per second.

Wheatstone's collected papers were published in 1879, and his portrait can be found in the National Portrait Gallery. He received many honorary degrees from Universities, became a Fellow of the Royal Society in 1836, and was knighted in 1868. He died in Paris in 1875. He suggested that his method could also be used to measure the speed of light, and that it could be used to tell whether light went faster in a medium such as water, supporting the Newtonian corpuscular theory, or slower, supporting the wave theory. These were the suggestions taken up by Arago in his paper of 1838.

Up to this point, everything suggested that, if not instantaneous, the transmission of light was nevertheless extremely rapid. And while huge astronomical distances facilitated the measurement of time, estimates of speed were hampered by the large errors in astronomical distance measurement, including the radius of the Earth's orbit around the Sun. To really nail down the speed of light, what was needed was a terrestrial measurement between two points on Earth a known distance apart. Many such measurements were made in the second half of the nineteenth century in Paris, when quantitative science flourished. The consequences of these measurements went well beyond the speed of light, because they led to an understanding of its importance for the classical theory of electromagnetism, as we shall see in Chapter 6. And while these experiments had originally been based in astronomy, their earthly context was one of political upheaval, national and civil war, and revolutionary developments in the arts.

Life in Paris in the nineteenth century was a turbulent succession of rival socialist and monarchist political factions. Following the French Revolution of 1789 and the time of Napoleon, King Louis XVIII was restored to the monarchy, then succeeded by his brother Charles X. In 1830 Charles was succeeded by Louis-Philippe, leader of the Orleanists, wealthy bourgeoisie who were willing to compromise with the revolutionaries. The 1848 "year of revolutions," in which revolts against authoritarian governments broke out throughout Europe, resulted in his fall and the start of the Second Republic and a new constitution, led by Charles Bonaparte (Napoleon III), a nephew of Napoleon.

Charles's regime ended when he declared war on Prussia in 1870, leading to his defeat and capture, followed by declaration of the Third Republic, which was to last until 1940. Paris was held under siege by the Prussians and starved into submission and surrender on January 28, 1871, after which the Prussians moved out of Paris.

There followed a kind of civil war during the time of the Paris Commune of 1871, which held power for two months. The commune attempted an ambitious program of reforms, including separation of Church and State, transfer of all Church property to the State, exclusion of religious instruction from schools, gender equality, and free education for all. Following bloody and bitter street fighting in Paris and the death and execution of thousands, power was eventually regained by moderate Republicans (today's liberals) into the time of the highly divisive Dreyfus affair of 1894, and leading up to the First World War in 1914.

In the arts, realism and modernism replaced romanticism, as writers such as Flaubert, Hugo, Balzac, and Zola brought French literature to international prominence. Chopin had reluctantly made Paris his home in 1831, Berlioz flourished soon after, Saint-Seans ("the French Mendelsohn") later. With impressionism and modernism came the highly original Debussy ("a composer whose music has no precedent") together with Faure, Ravel, Satie, Massenet, Stravinsky, and Rachmaninov, and others during the Belle Epoque (1872–1913). A steady stream of visiting artists—Liszt, Wagner, Verdi, Offenbach—confirmed Paris's standing as the center of European culture, as did the rise of the impressionist movement in painting (Monet, Renoir, Sisley, Manet, Pissaro, Cezzanne, Degas) from about 1860. In the sciences, Louis Pasteur had found a vaccine for rabies, and established the germ theory of disease with the heating process which killed bacteria and led to pasteurization of milk, while Fourier, Carnot, Becquerel, and Poincaré added lustre to Paris's reputation as the European center of scientific enquiry.

Paris had been largely demolished and rebuilt under Napoleon III in the 1850s and 1860s by Haussmann, producing the wide avenues, statues, and many of the beautiful buildings we see today. Conditions were a vast improvement on the time of the French Revolution, but still without electricity, proper drainage, clean water supply to most houses, or convenient heating. Horse-drawn carriages and manure filled the streets, but steam trains were rapidly being deployed. Photography had progressed from the primitive Daguerre process to the wet plate collodion method of 1851, producing excellent monotone images such as that shown in **Figure 5.9** of the multi-story buildings which were subject to the Prussian bombardment in 1871, when messages were sent out of the encircled Paris by balloon and returned by homing pigeon. It was against this background that Marie Cornu's measurements of the speed of light took place, as we discuss later, running beams of light across the rooftops and between the chimneys of Paris.

It was Francois Arago (1786–1853) who first took up the challenge of terrestrial measurements of the speed of light, and he based his first studies on the earlier ideas of Wheatstone that we have described. Arago had a most adventurous life, living through many historic events, making scientific discoveries, and having considerable political and social influence. Educated at the Ecole Polytechnique, the story is told that Napoleon Bonaparte requested in 1803 that all students sign a petition supporting his appointment as Emperor. Francois refused, to which Napoleon, on noting that he came top of the class responded "One can't send down the top student. If only he'd been at the bottom…" Soon after, Poisson appointed him secretary to the Paris Observatory. With Biot, he was sent to Spain to map out a meridian arc, in order to determine the length of the meter, defined after the French Revolution as one ten-millionth of the distance from the Equator to the North Pole. Unfortunately his surveying activities were misunderstood by the local population as those of a spy for a French invasion, and the twenty-two year old was imprisoned in the Bellver fortress in 1808. He soon escaped in a fishing boat to Algiers, but was once again captured by pirates and imprisoned at Palamos. After release and further adventures, including a trek along the North African coast from Bougie to Algiers, he reached Paris with his meridian notes intact, which he was able to deposit at the Bureau de Longitudes. Around this time he became a close friend of Alexander von Humbolt. He was rewarded by election to the French Academy of Sciences, appointed to a chair of analytical geometry and, as an astronomer of the Paris Observatory given a residence there for the rest of his life.

Arago's scientific research covered a wide field, including the discovery of rotary magnetism and eddy currents. Most famously, as we have described, he displayed the bright spot in the center of the shadow of a disk for his friend Augustin Fresnel, known ever since as "Arago's bright spot." He received many awards, including the Copley medal of the Royal Society. As a scientist, he was described by one contemporary as "like a hurried explorer who travels through virgin lands, gives them a name, then rushes on toward more distant horizons." As a member of the Chamber of Deputies, and secretary of the Academy of Sciences, he was influential in obtaining a prize for Daguerre for his invention of photography and a grant for publication of the works of Fermat and LaPlace. He was active in the development of the railways and telegraph system, gave a series of public lectures on astronomy for nearly thirty-five years, and provided an invaluable set of memoirs of deceased Academy members. His "rapidity and facility of thought, his happy piquancy of style, and his extensive knowledge peculiarly adapted him to the position he was given as perpetual secretary of the Academy in 1830" said a colleague.

Arago was a dedicated liberal Republican (well to the left wing of the politics of those days), consistently so through the changing regimes of France at that time. With the fall of Louis-Philipe he joined the provisional government in 1848, the "year of revolutions," becoming Minister of War and also of Marines and Colonies. In these positions he managed to improve rations and abolish flogging in the navy, and to abolish slavery in the French colonies. The new government of Louis Napoleon in 1852 again required an oath of loyalty from all its office-holders, which he refused to give. However this was overlooked, and he died from diabetes the following year.

For the purposes of our story, Arago is particularly important for his 1838 paper, in which he proposed adapting Wheatstone's method to compare the speed of light in air with that in water, and thereby addressing the question of whether it is a wave (if it slows down in water) or a particle (if it speeds up, as they incorrectly believed at the time. **Figure 5.3** shows the principle of the method. The apparatus,

using an early application of helical gears, was built by his gifted technician Louis Breguet, grandson of the famous watch-maker Abraham
Breguet, who also built the rotating mirror system for Marie Cornu's
later measurements in 1876. The beam passing through the water in
Figure 5.3 is at a different height from the second beam. As drawn,
going slower, the retarded light pulse (from a spark) arrives after the
rotating mirror has made a small rotation, and so is sent off in a slightly
different direction from the earlier pulse which did not traverse the
water. The experimenter aims to measure this divergence angle θ
shown between the two beams of light. By turning the beams on and
off, the experimenter could also tell whether the light speed had
increased or decreased when passing through the water, and could
measure its speed. Arago discusses the experimental parameters in
detail. He found that an easily achieved rotation speed of about 1000 RPM
is sufficient, that repeating electrical sparks are brief enough to allow θ
to be measured, and that a tube of about twenty-eight meters filled
with water would be required. This length could make the spark too
dim, a situation improved by using a higher mirror rotation speed.
Arago built this apparatus and spent a decade perfecting it (including
the introduction of multiple reflections, and evacuating the system to
achieve higher mirror speed by reducing drag). However his eyesight
failed before he could obtain a result. In reality, this design could never
have worked for several reasons, including the need to synchronize the
sparks at the source with each other and with the mirror position to
prevent the flashes all emerging in different directions. But this design
inspired the two instruments which did work, beautifully: those of
Foucault and Fizeau, both born in 1819, a generation after Fresnel,
Young, and Arago.

Hippolyte Fizeau decided not to follow his father, a Professor of
Medicine at Paris University, into the same field, apparently because he
suffered from migraine headaches, but instead taught himself physics.
He was lucky to be taken on by Arago at the Paris Observatory, where
he worked on the improvement of Daguerreotype photography, for
applications to astronomy. This was to have a huge impact, allowing
stellar positions to be recorded and measured with ease, rather than
sketched using pencil and paper. Fizeau collaborated with Foucault in
taking the first successful Daguerreotypes of the Sun.

In 1848 he published a correct analysis of the "Doppler effect" for
both sound and light, unaware of Christian Doppler's earlier 1842

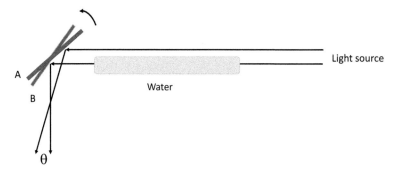

Figure 5.3 Arago's scheme for comparing the speed of light in air and water. The two beams are at different heights above the page. By the time a light pulse arrives at the mirror after being slowed by passage through water, the rotating mirror at A has turned a little to B, deflecting the beam into a different direction from the un-delayed beam. The experimenter measures this angle θ, and from the known rotation speed of the mirror can calculate the delay.

analysis. This important effect accounts for the fact that the pitch of a train whistle increases as it approaches and decreases as it recedes—in other words Doppler's formula gives the observed frequency of a moving source of waves, showing how it depends on the speed of the source or detector. (Think of a speedboat running into water waves—the crests hit the boat more frequently than if the boat were stationary.) However, Doppler had assumed incorrectly that the spectral colors in starlight were due to his "Doppler effect" acting on very high stellar velocities, rather than the emission from different atoms actually responsible. Fizeau drew the more useful conclusion that a much smaller relative velocity between stars and the Earth could be measured this way, since one could compare the emission from a certain type of atom in the star with the emission from the same stationary atom on Earth. This *redshift* technique rapidly became one of the most important in astronomy and was used crucially by Hubble and others to establish the expanding universe model in the 1920s, and hence provide evidence for the Big Bang theory. In a second collaboration with his friend Foucault, they achieved a record for interferometry by observing interference effects over a path difference of 7000 wavelengths of light. Fizeau designed two famous interferometers—one is still used today to test for optical flatness in optical components, the other was used to test the Aether drag hypothesis, as we will discuss later.

In 1849, Fizeau, after ending his collaboration with Foucault, reported the first terrestrial measurement of the speed of light. In spite of becoming competitors thereafter, they had behaved with impeccable propriety toward one another up to that point, giving the fullest credit to each other and to others who had contributed ideas, such as Herschel, Arago, Wheatstone, and Bessel. Rather than adopt his mentor Arago's proposal, he decided to use a toothed wheel to interrupt light reflected from a mirror, as shown in **Figures 5.4 and 5.5**. His technician Paul Froment built this toothed-wheel apparatus (and also the rotating mirrors and pendulum for Foucault). Froment, who had worked in Manchester, was said to be able to split hairs into four pieces lengthwise, and drill a hole down a darning needle. Fizeau commented that he could obtain ivory combs with teeth a quarter of a millimeter wide. Fizeau's use of a wheel with 720 teeth greatly increased the time resolution over Arago's arrangement, while giving a quantitative result and solving the synchronization problem, since the light is broken up into pulses by the wheel, which also helps in its detection.

In **Figure 5.4**, if the time taken by light to travel from the source at S, through a gap between the teeth, to the mirror and back, is just equal to the time for which the toothed wheel rotates by 360/720 degrees, around to the next gap between the teeth, then the observer at D will see bright light. (The observer is able to see through the half-silvered mirror shown dashed, which both reflects and transmits light.) Light travels about a foot every thousandth of a microsecond (a millionth of a second), so that with a thousand teeth on Fizeau's wheel, rotating at 1000 revolutions per second (easily achieved using modern equipment), one would only need a thousand feet of delay, or half this distance to a mirror, for the light to be reflected to the next tooth. If the speed of the wheel is increased, the return path will become blocked when light from S passes through a gap between the teeth, but returns to be obstructed by a tooth. Thus the time for the light transit can be determined from the rotation speed of the wheel.

Fizeau established his light source in his father's house at Suresnes near Mont Valerien, but the mirror was placed at Montmatre, a distance of 8.66 km (5.4 miles) away. The light source was an ether flame on lime, and the experiments were conducted in the evening. The mirror rotation was driven by clockwork, powered by a descending weight—a rotation speed of about 12 revolutions per second was sufficient. His result, in 1849, for the speed of light, was 3.14×10^8 m/s, against the modern value of

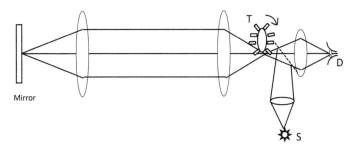

Figure 5.4 Fizeau's apparatus for measuring the speed of light, showing rotating toothed wheel T, source of light at S, and detector at D. The light leaves between the teeth, but by the time it comes back from the mirror a tooth has moved around to block it.

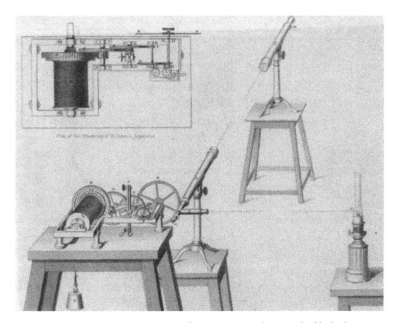

Figure 5.5 Fizeau's 1849 apparatus for measuring the speed of light between his father's house in Suresnes and Montmatre. (From Tobin (1993).)

2.99×10^8, an error of about 5%, but only slightly larger than the astronomical measurements of the time. The result won him the Triennial Prize created by Napoleon III for 30,000 francs, or six times the annual salary of the physicist position occupied by Foucault at the Paris Observatory. The weakness of the method is the difficulty of judging exactly when the light is eclipsed, a problem later solved by Marie Cornu in his much more accurate measurement of 1876 using the same method. He obtained a value of 300,330 km/s. Fizeau himself then spent many years attempting to improve his apparatus, however Arago, a great supporter of the project, died in 1853, and Fizeau's technician Froment died in 1865, before completing the new apparatus. Fizeau also worked on the measurement of the speed of electricity, publishing his results in 1850, and on the measurement of the speed of light in water (compared with air) using rotating mirrors, in competition with Foucault.

A moody and reserved man, in 1853, Fizeau married Theresa de Jussieu, the daughter of a famous botanist, with whom he had three children. His withdrawal increased on the death of his wife in 1866. He became a member of the Academy of Sciences, was awarded the Rumford Medal of the Royal Society of London, and is one of the seventy-two scientists whose names are inscribed on the base of the Eiffel tower. He was the only one of these still alive when it was opened for the World's Fair in 1889. After his death in 1896, the secretary of the Paris Academy said of him, "Those, of whom there are only a few now, who knew him during his last years, can remember a venerable old man with a shock of hair and a thick beard, whose behavior was impressive, though rather cold. The interests of science alone made him forget his habitual reserve; although Fizeau did not like arguments, he became an adversary not to be disregarded whenever there was a discussion."

Leon Foucault's method (published in 1862), which he spent twelve years perfecting, was different from and more accurate than Fizeau's. Foucault realized that it would be more accurate to measure the angular deflection of light from a rotating mirror directly while the light was travelling to and from a distant mirror, much as Arago had suggested, but with a clever modification which solved the synchronization problem. According to Cornu, the collaboration between Fizeau and Foucault ended on the day in 1849 when they both realized, in conversation, that all the problems with Arago's original experiment comparing the speed of light in water and air could be solved by sending the light back on its tracks, after reflection from a fixed mirror, to the rotating mirror. By the time the light got back to the rotating mirror, it would have rotated slightly, sending the beam off in a different direction. Cornu later described their efforts from that time on as "a steeple chase." Foucault and Froment worked on rotating mirrors at his mother's house, while Fizeau and Breguet worked on toothed wheels in the

Meridian room of the Observatory. The race was on. Both had the blessing of their mentor Arago, but it does seem that he favored Fizeau.

Figure 5.6 shows Foucault's arrangement for his later measurement of the speed of light. The reflection of light from a mirror is known as *specular*, meaning that incident and reflected rays make equal angles with a line drawn at right angles to the surface of the mirror. Because of this, we see an image of ourselves in a mirror the right way up (erect), but left-to-right inverted, that is, with our watch (for those, unlike my students, who still use them!) on the wrong wrist (right instead of left) in the mirror image. In **Figure 5.6**, the lens L focuses a light source S1 through the rotating mirror M1 (which rotates continuously through 360°) onto the surface of the curved mirror M2 as a small spot. As this spot sweeps across M2 it chops up the reflected light (which is only present when the beam is on the mirror) into short pulses. By the time a light pulse has returned from M2 to M1, the mirror M1 has turned slightly, sending the returning light back to a displaced focus through lens L at S_2. Foucault would measure the distance x between the two focii S_1 and S_2 (about one millimeter) using a microscope. In his initial experiments he actually used a half-silvered mirror inclined at 45° to the beam to view both spots from the side. More details of this clever arrangement are given in Appendix 3, showing why the source image does not move when the mirror is rotated slowly—devising such an optical arrangement was the crucial breakthrough which made the

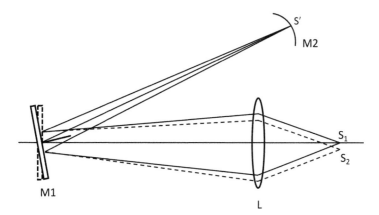

Figure 5.6 Foucault's rotating mirror system for measuring the speed of light. The distance between S_1 and S_2 is x. This distance increases as the mirror speed is increased.

experiment work, improving greatly on Wheatstone and Arago's method.

Foucault used focused sunlight as his source, with a clockwork mechanism to keep the source stationary as the Earth turned (a heliostat). He used a high-pressure air bellows to drive the rotating mirror, shown in **Figure 5.**7, built by his friend the famous organ builder Cavaille-Coll, who had built the Notre-Dame pipe organ. With a regulated pressure stability of one part in 1500, the speed could now be measured by a new stroboscopic method involving a toothed wheel, which could be viewed through the microscope simultaneously with the light spot deflection. The light path was extended and folded using five concave mirrors, for a total length of twenty meters. His 1862 result for the speed of light is very close to the value we use today; he obtained 298,000 $+/-$ 500 km/s, the high accuracy mainly due to his stable mirror rotation speed and independent method of measurement. Unlike Arago's proposal, which could produce a lighthouse beam sweeping around a room, Foucault had used the finite width of his curved mirror to break the reflected light up into pulses. During this pulse, as the light spot focused onto the surface of the curved mirror sweeps across it, light is continuously reflected by the rotating mirror, which therefore returns it to the detector over a range of angles, broadening the spot and limiting the time resolution of the instrument. With the speed of light c, if the time light takes to travel the distance 2D to M2 and back is t, then ct = 2D, while in this time the mirror, rotating at n revolutions per second, has turned through an angle 2 n π t. In terms of the dimensions of the apparatus shown in **Figure 5.6**, the speed of light is then

$$c = 8\pi n \, a \, D^2 / \left[x(b + D) \right].$$

A graph of x against n will be a straight line whose slope gives the speed of light c. His result (1862) was proclaimed as a great achievement, and interpreted by the physicist Jacques Babinet as "the accurate determination of the distance to the Sun" by laboratory methods.

Foucault also undertook important measurements of the speed of light in a refractive medium—recall that the Newtonian corpusculists believed it should travel faster in a denser medium. His clever experimental arrangement for his competition with Fizeau in 1849–50, similar to **Figure 5.6**, allowed light to be sent from the rotating mirror (driven by a steam turbine siren) successively through air to mirror M2 or, on

Figure 5.7 Foucault's 1862 rotating mirror, driven by compressed air from a pipe-organ pump. (From Tobin (1993).)

the other side of the apparatus, through a tube filled with water. In this way the speed of light in the two media could be compared by observing spot shifts directly. His result favored the wave theory of light, with light slowing down in the water. He published this result slightly before Fizeau on February 17, 1850, in a Paris newspaper (*Journal des Debats*). On May 6, the journal *Academie* published papers by both Foucault and Fizeau, Foucault providing his result, supporting a wave theory and apparently demolishing the particle theory, while Fizeau reported that "the state of the atmosphere has not permitted us to make any observations...if our experiments are not completed it is because we waited before beginning them for M. Arago to authorize us to embark on a topic of research which belonged to him." Clearly a touchy subject! Fizeau finally reported success, in agreement with Foucault, on June 17, 1850. Foucault submitted these results for his Doctor of Science (loosely PhD) degree. He scraped through.

But Foucault is far more famous for another demonstration of his, when in February 1851 he erected a huge pendulum in the Paris Observatory, and later in the dome of the Pantheon in Paris. The arc of this pendulum, a straight line traced out beneath the bob on the floor, ran in successively different directions across the floor (through the central point of the pendulum's swing) throughout the day. This straight line can be made, for example, by allowing fine sand to fall out of the bob onto the floor as it swings. After an hour the sand traces out two fan shapes on the floor, due to the Earth's rotation beneath the pendulum, with the fan points meeting at the center. Parisians were amazed to see this vivid demonstration of the rotation of the Earth, some attributing it to witchcraft. It is easy to understand how this works if we imagine the pendulum to be erected at the North Pole. The plane of the pendulum's swing remains fixed, defined by its starting condition (not by the fixed remote stars!), while the Earth rotates beneath it. The constant direction of the pendulum's swing (in spite of the Earth moving below it) is a vivid demonstration of one of Newton's laws about bodies continuing in their motion unless disturbed by external forces. To keep the pendulum swinging permanently, it must be driven by a motor, which must not bias the motion.

You can find many modern and fairly simple set-ups for repeating Foucault's speed of light (or pendulum) experiments from student laboratory notes on the web. Electronically controlled motors (with octagonal mirrors) rotating at speeds over 1000 Hz are readily available now from used bar-code readers or laser printers, which operate on similar principles, and a more sophisticated implementation using modern digital electronics is described in the research paper by Vörös and Weihs (2015).

Unlike Fizeau, Foucault never had much use for equations in his work—his great gift was design and construction of ingenious precision instrumentation, and for this lack of theoretical analysis, his work was initially discounted by some. He was, however, awarded the Copley medal of the Royal Society of London and eventually elected a member of the French Academy of Sciences. He went on to invent new designs for gyroscopes and speed governors, a new design of photometer, and discovered induction currents due to AC fields. He died in 1868. His life story is beautifully told in William Tobin's book.

These mechanical methods of measuring the speed of light were to be perfected in the following decades by several physicists, including Newcomb (in 1881 in Washington, using the rotating mirror method), and Michelson in 1927. Marie Cornu (1841–1902), an astronomer at the Observatory, made the last and most accurate measurements in Paris by the rotating mirror method. He was motivated both by developments in the field of electrodynamics, a field entirely distinct from astronomy, and by the imminent Transit of Venus (December 9, 1874), which would hopefully give a more accurate value of the Earth–Sun distance. Based solely on theoretical considerations in electrodynamics, James Clerk Maxwell had predicted in 1868 a value for the speed of light in terms of electrical constants, having apparently nothing whatsoever to do with light. It was urgent to test this idea, which was to have a profound influence on Einstein's theory.

Marie is shown at work in **Figure 5.8**, in 1874, only three years after the Siege of Paris. The Siege included an artillery bombardment of Paris by Bismark's forces, which resulted in the surrender of the garrison, and the occupation of Paris by German forces in 1871. **Figure 5.9**, using the newly invented wet-plate collodion photographic process (1851), shows the state of some Paris buildings after the attack, which caused more destruction to the city than any other war before or since. The bombardment also led to starvation of the population under siege. **Figure 5.10** shows a restaurant menu from that time, offering fricassee of rats and mice to customers, which, as noted, included the American Ambassador to France, Elihu Washbourne. He was the only representative

Figure 5.8 Cornu at work in 1874, casting beams of light across the rooftops of Paris, reflected by distant mirrors to allow him to measure the speed of light. Chimney smoke from evening cooking was his biggest problem. (From Tobin (1993).)

Figure 5.9 St Cloud, Paris 1871 during the Siege of Paris in the Franco-Prussian war, showing building damage and the high quality of photography achieved by that time. (From Wikipedia.)

of a major power to remain throughout the siege, and assisted with humanitarian relief and communication between the combatants.

Cornu started out as a mineralogist and became a professor of physics at Ecole Polytechnique, supported by Fizeau. He did much research in spectroscopy, measuring Doppler shifts on the Zeeman effect on the sodium D optical emission spectrum of light from the Sun. He also made new measurements of the constant in Newton's law of gravity, using a miniaturized version of Cavendish's apparatus with a fine torsional fiber balance. He made several hundred speed of light measurements, using both rotating mirrors, and improving on Fizeau's method, in which the major error arose from difficulty in judging the exact moment of darkening, as the toothed wheel rotated and its speed was varied. Cornu was able to automate this process using electrical contacts to measure the rotation speed using a 10 km light path. He used a simple carriage clockwork motor without escapement to provide up to 750 revolutions per second. Heat from chimneys in the light path

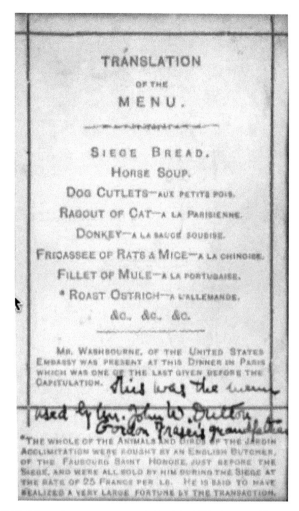

Figure 5.10 Menu from a restaurant in Paris in 1871, showing Fricassee of Rats and Mice, offered during the siege of Paris, and attended by the American Ambassador. (From Google Images.)

caused trouble, so he waited for the school holidays for his best measurements. Without the collimated intensity of a modern laser pointer, all of these experiments were plagued by the weak light sources of the day, and the tendency of the light to spread out—there was a constant battle to obtain larger and better quality collimating lenses and mirrors to confine the light over long distances; and to measure these distances across Parisian streets accurately by triangulation. Owls nesting in the return mirror collimator tube at the Tour de Montlhery fort, which sent the light back to the Observatory, were also a problem. The best results were obtained in September 1874, giving the value 300,400 km s^{-1}, with an accuracy of one part in one thousand. For this he was awarded the 10,000 franc Prix Lacaze scientific prize in physics for 1877.

For the next hundred years, Cornu was far more famous to generations of physics students (including myself) for his "Spiral," a graphical construction for evaluating Fresnel's near-field diffraction integrals before the days of desktop computers—it was still quicker to use than submitting a program to a mainframe computer with one-day turnaround in the 1960s and 1970s.

Following Cornu, the young Alfred Michelson at the US Naval Academy in Annapolis used the rotating mirror method in 1879 for light-speed measurements, with a tuning fork linked to the mirror to maintain a steady speed. (Recall that notes around the middle of a piano have frequencies of around 400 Hz, which is fast enough for the mirror rotation speed.) He returned to this method (using polygonal mirrors) many years later in the nineteen twenties for the last use of the rotating mirror approach, and died during this experiment in 1931.

All of these scientists who devoted their lives to measurement of the speed of light and to understanding its properties would have been astonished by the latest developments in this field. For example, it has recently been proven possible to take a movie of a pulse of light as is passes by at the speed of light, as demonstrated by Ramesh Raskar and his colleagues in their laboratory at Massachusetts Institute of Technology (Velten (2013)). Light travels about one foot every nanosecond, and Raskar's very fast camera has an effective frame speed of about two picoseconds per frame. A nanosecond is a thousandth of a millionth of a second, and a picosecond is a thousandth of a nanosecond. The short light pulse therefore travels about half a millimeter between each frame. The camera detects light scattered from the pulse by dust. Using a kind of "streak" camera, this group can build up a movie showing the blob of light about a centimeter long running along the axis of a bottle at the speed of light.

As the astrophysicist W. Tobin has pointed out, the work of these sci entists tells us a lot about how science was conducted in the nineteenth century as it changed toward our modern culture. At that time, the *pli cachete* was a sealed document which could be deposited with the Academy at any time, describing a scientist's newest idea. This would protect his or her claim to priority, in case someone else later came up with the same idea. Fizeau had deposited his in 1849 for the toothed-wheel idea (finally opened in 1982!). In this way, scientists could *own* an idea, even if they were not working on it, and so prevent others from doing so. It is clear that Arago gave Fizeau his blessing to work on his ideas, but for Foucault, whose quaint personality was described as like "a pasha with three pigtails," it was more difficult, and a matter "of extreme delicacy" for him to ask Arago for permission to work on speed-of-light measurement, thus ending his collaboration with Fizeau, according to his colleague Lissajous. The Observatory director Le Verrier (discoverer of the planet Neptune) who followed Arago and supported Foucault, argued against the *pli cachete*, describing it as "an eternal menace suspended over the heads of those working in a new field. Proprietorship, applied to ideas, to *intentions*, is an essentially modern phenomenon which is a sign of intellectual poverty rather than wealth." Le Verrier's extraordinary achievement, among many, was to provide the first complete mathematical analysis, unaided by computer, of the perturbing effect of a second planet on the motion of the Earth around the Sun—the notoriously difficult "three-body problem" of classical physics. But it is only in the work of these two, Le Verrier and Cornu, at the end of the century that we see the modern approach to research emerge, with their long and detailed published accounts of their work, clearly intended to provide enough information for others to repeat and confirm, as all students in research are now taught to do.

The modern convention is that all scientists are free to work on the published results of other researchers; in astronomy and structural biology in particular large international databases have been established, accessed via the web, to store results in strictly controlled formats. The researchers who first obtained the data have a fixed period, perhaps a year, between announcing their discovery and providing public access to their data, during which time they alone can use the data for their own purposes, in order to establish priority in its interpretation and discovery. The independent database custodians may then become a kind of "archive police," who may also have access to the very latest theories and data analysis algorithms (including now those based on machine learning), which were not invented when the data was deposited years ago. We thus have the remarkable (and desirable) situation where an elderly scientist who deposited data long ago with supporting theoretical interpretation may suddenly find it republished by others in support of a completely new and different scientific theory.

The French results on the speed of light also resulted in important theoretical advances from Lord Rayleigh and others in discussing what exactly these methods were measuring. Rayleigh and others realized that there were two different kinds of "velocity of light"—which came to be known as the phase and the group velocity, which we will discuss in Chapter 6. These ideas have helped us to clarify many schemes which appear to promise faster-than-light communication.

By the end of the nineteenth century, the wave theory of light was triumphant. Five years later, Einstein, building on Planck's work on black-body radiation, would publish a paper on the photoelectric effect, which earned him a Nobel Prize, for showing that light was a particle. In a similar way, the physicist J.J. Thomson in 1897 showed that the electron (which, unlike the photon, has mass) was a particle. His son, G.P. Thomson (with others) showed in 1927 that it was a wave. All four received Nobel Prizes for their efforts. This wave–particle duality lay at the heart of the foundation of quantum mechanics during the 1920s. Modern physicists accept this duality, with the idea that quantum particles "travel as waves and arrive as particles," while continuing the search for deeper understanding and a better theory. The background to all this wave–particle duality lay with measurements of the speed of light. As the science historian G. Sarton has said, "the story of the Arago–Fizeau–Foucault experiments will always be one of the most beautiful in the history of mankind."

6

Faraday and Maxwell
The Grand Synthesis

We must now launch two completely independent threads to our story—a very brief history of electrostatics and of electromagnetism, followed by their unification and the resulting emergence of a comprehensive theory of light and radio waves. From this followed the theory of the electromagnetic spectrum, including radiation at wavelengths longer than light. And it has brought us the wonders of radio, television, the Internet, satellite communication, and the mobile phone, all communicating at the speed of light, and radiation at wavelengths shorter than light, such as X-rays used for imaging in medicine. Along the way and by the end of the nineteenth century, scientists came to understand how radio waves can propagate through space even if, as Einstein showed in 1905, this space is truly devoid of even the mysterious Aether.

Electrostatics is the study of the effects of electrical charges and the forces they produce; electromagnetics that of magnets and the forces they produce. We are all familiar with static electricity: electrical charges which give you an electrical shock on the doorknob of a hotel room if you are wearing rubber soled shoes and walking on dry carpet. An electrical current is just a flow of these electrical charges, usually electrons flowing along a wire.

Prior to 1800, the effects of magnets and charges were believed to be entirely unrelated. Their unification around 1830 led on to the fundamental theory of the propagation of light, in vacuum or media, exactly as had been anticipated by Huygens and Fresnel. The towering figure in this mighty intellectual achievement was the Scot, James Clerk Maxwell, whose equations remain central to modern physics. Most physicists would agree that his equations, plus a few from Einstein, Newton, and Dirac, are the greatest in all of physics. Einstein himself once describe Maxwell's work as the "most profound and the most

Lightspeed: The ghostly Aether and the race to measure the speed of light. John C. H. Spence.
© John C. H. Spence 2020. Published in 2020 by Oxford University Press.
DOI: 10.1093/oso/9780198841968.001.0001

fruitful that physics has experienced since the time of Newton." In a profoundly mysterious metaphorical process, Maxwell managed to obtain the correct equations to describe the propagation of light and radio waves in vacuum using a mechanical model, based on wave propagation in a special kind of invisible very stiff ultralight rubber—the Aether, which was supposed to fill the universe. This could also support the flow of fluids along tubes, and whirling vortices! On completion of his theory near Edinburgh in 1864, showing that light was an oscillation in an electromagnetic field, he found that it made an independent prediction for the speed of light which agreed "pretty much" with the measured value. In his letters he describes this confirmation as the most exciting day in his life.

The connection between electricity and light was suggested by the fact that electrical sparks produce light, especially from lightning, as Ben Franklin found by flying a kite into storm clouds and collecting the charge. Another direct connection between electromagnetics and light was provided when Michael Faraday (1791–1867) discovered the magneto-optical effect in 1846. Here the plane of polarization of light is rotated if it passes through a material subject to a magnetic field. Fresnel and Arago had established in 1816 that, assuming light were a wave, then its oscillations occurred transversely to the direction of propagation like an ocean wave, not longitudinally like sound waves.

Prior to the nineteenth century, no connection was understood between the forces produced by magnetic fields and those from electrical charges. A "field" is simply a force experienced by a very small charge or magnet at some point due to charges and magnets at other positions. Charges could only be generated by friction prior to the invention of the battery by Galvani and later Volta in 1779; magnets had been known since antiquity and used for navigation in compasses, but were first studied extensively by Gilbert around 1600. Gilbert made an extensive comparison of magnetic and electrostatic forces, pointing out that the Earth was itself a gigantic magnet, with north and south poles. In 1785, Charles Coulomb established that the force between charges varies inversely as the square of the distance between them.

The unification of electricity and magnetism began with the work of Oersted, Ampere, and Faraday. In 1820, Oersted made the crucial discovery that a current passing through a wire deflected a compass needle, suggesting that the charges moving through the wire somehow creat a magnetic field outside it, affecting the compass. Ampere, hearing of this discovery, found that a coil of wire carrying a current behaved like an ordinary bar magnet, suggesting its use for sending code by telegraph. He went on to quantify the force between two current-carrying wires due to their fields acting on each other, providing a mathematical description now known as Ampere's law. This led to the invention of the galvanometer, the first instrument to reliably measure current by deflecting a pointer on a scale.

But perhaps most important of all, was Faraday's discovery in 1831 of electromagnetic induction, in which a *changing* current in one coil produces a current in another nearby. Equally important, he found that moving a bar magnet through a coil of wire produced a current in it, eventually giving us the electric motor, dynamo, and generator. The later quantification of this effect became one of Maxwell's famous four equations, while the others were based on Coulomb's and Ampere's laws. Faraday himself constructed a simple electric motor and the first dynamo, thereby originating the electrical power generation industry.

Born in 1791 to a poor family with very limited education, Faraday would certainly never have been considered "a gentleman" in the rigid class system of the day. As a teenage apprentice to a book-binder he managed to read widely and provide himself with a rudimentary education, learning about electricity from the Encyclopedia Britannica at his workplace. He was devoutly committed to the Sandemanian religious sect, which was related to the Church of Scotland, in which he became an elder, marrying Sarah Bernard, whom he met through the church in 1821. They had no children.

The story is famously told of how Faraday at the age of twenty started to attend lectures by the famous chemist Humphry Davy at the Royal Institution, which Faraday wrote up and presented to Davy, whereupon Davy appointed him as his assistant. Faraday toured European scientific centers with Davy in 1814 as Davy's valet, badly treated by Davy's wife as a servant, but intensely curious about the new science he was seeing at these centers. He died at the age of seventy-five, after receiving many honors, including election as a Fellow of the Royal Society in 1824 and being awarded many honorary degrees, but he turned down a Knighthood on religious grounds,. He became the first Fullerian Professor of Chemistry at the Royal Institution in 1833, a position created for him by a bequest from the eccentric John Fuller, an English Member of Parliament, and he twice turned down the Presidency of the Royal Society. He was elected an Honorary Member

Figure 6.1 Faraday in his laboratory in around 1835, at the Royal Institution. (From Jones (1870).)

of the American Academy of Arts and Sciences, and given free lodgings in Hampton Court by Prince Albert in 1848, where he died in 1867. Faraday is shown in his laboratory in **Figure 6.1** around 1835.

Perhaps the greatest experimental physicist of the nineteenth century, he made many discoveries in addition to electrical induction and the magneto-optical (Faraday) effect. Using the newly invented battery, he established the laws of electrolysis (the basis of electroplating), introducing the terms "anode," "cathode," "ion," and "electrode". He was responsible for many other discoveries in chemistry, such as the discovery of benzene, the liquefaction of chlorine, and diamagnetism. He invented the Faraday cage, a metal box within which electrical fields are absent. More than anyone he was responsible for the later popular use of electricity.

His great importance to the story of the measurement of the speed of light is his idea of "fields of force." An electric field is a force on a small charge at some point, and he supposed these to fill all space and, as we now know, travel at the speed of light. (Strictly, the electric field is the force per unit charge at a point.) It was the mathematical formulation of Faraday's physical intuition by Maxwell which finally put the wave theory of light on a firm theoretical foundation, and gave the speed of light a particular value solely in terms of other fundamental constants.

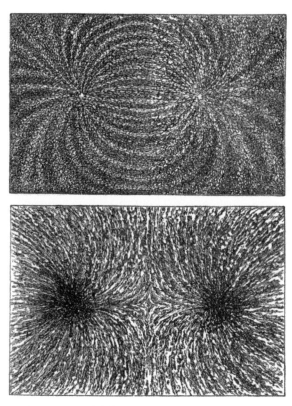

Figure 6.2   The magnetic field lines of force seen by Faraday, which gave him the idea that radiation would result if they vibrated. He passed the idea on to Maxwell to express in mathematical form. Small iron particles on a horizontal plate are seen lining up in the field running between the poles of a horseshoe magnet below. Upper: north and south poles. Lower: two north poles. (From Lodge (1889).)

Faraday had presented his idea of field lines at the Royal Institution following a lecture by Wheatstone which had ended prematurely. In a letter to a friend, he suggested that his lines of force (shown in **Figure 6.2**) were like fine invisible curved elastic bands connecting charges or magnets, "with equal pressure in all directions at right angles to these lines." He "considered radiation as a high species of vibration in the lines of force which are known to connect particles and also masses of matter together." This was the idea which inspired Maxwell to develop his field equations. The fields are themselves derived from more fundamental quantities called potentials, which Maxwell used, and which also travel at the speed of light throughout space. In the period after Maxwell's death, due partly to Heaviside's work, the focus moved to forces and fields, which were then considered fundamental and more useful.

The foregoing experimental work provided the essential background to Maxwell's remarkable theory, whose development after his death is compellingly described in detail in Bruce Hunt's book, *The Maxwellians*. There, and in Darrigol's superb history of optics, you can find the remarkable account of how Maxwell used a mechanical model of the elastic, invisible Aether medium to develop his famous equations. The same equations turned out to apply in complete vacuum, where they correctly describe the propagation of light. Maxwell took the existence of the Aether for granted at the outset, and set himself the problem of finding the equations that described the coupled time dependence of electric and magnetic fields. The "fields," postulated by Faraday, are *lines of force* which emanate from charges and magnets. They were suggested to Faraday by observing iron particles, shaped like grains of rice, on a horizontal card placed across the poles of a horseshoe magnet, as shown in **Figure 6.2**. The grains form curved lines running from one pole (north) to the other (south). Faraday imagined that these invisible *field lines*, filling space, would exert a force on any small charges or magnetic poles which they crossed, producing a kind of tension throughout the vacuum of space. This was a completely different idea to the prevailing instantaneous "action-at-a-distance" basis for Newtonian mechanics, and we now know that these fields and associated forces travel at the speed of light.

Faraday wrote to Maxwell with this idea, who expressed it in mathematical form, leading ultimately to today's Field Theory in physics. Theories of "delayed action at a distance" were developed by several scientists in the late nineteenth century, producing various paradoxes of causality. Maxwell even considered the possibility of perpetual motion. Faraday's preoccupation with the idea that magnetic and electric forces take time to propagate can be seen in his diary from near the end of his life, when he made a futile attempt to observe the delay (rather like Galileo) between switching on a magnet and its effect, across the backyard of the Royal Institution in London, by watching the needle of a galvanometer. He wrote,

It would appear very hopeless to observe the time in magnetic action if it at all approached to the time of light, which is about 190,000 miles in a second, or that of electricity in a copper wire, which approximates to the former. But then powers which act on interposed media are known to vary, and sometimes wonderfully. Thus the time of action at a distance is wonderfully different for electricity in copper, water and wax. Nor is it likely that the paramagnetic body of oxygen can exist in air and not retard the transmission of the magnetism. At least . . . such is my hope

In a way he was right—recent research has produced media with very high dispersion, in which the speed of light (the group velocity) can be reduced to walking speed.

Once again we should step back and see here how scientific advances actually occur in practice, in this crucial matter of whether Newton was right, that gravitational and electrical forces act instantaneously across the universe, or whether they propagate at the speed of light. Faraday's physical insight (based on **Figure 6.2**) gave strong hints of the answer before Maxwell's theory took over, leading to the overthrow of Newton's authoritative instantaneous *action at a distance* concept, and leading eventually to our current theory for light, radio, and gravity wave propagation at the speed of light. The interplay between experiment and theory was crucial. Maxwell retained a very high regard for Faraday's physical intuition, despite Faraday's very limited use of mathematics, which did not even extend to trigonometry. Einstein himself kept pictures of Newton, Faraday, and Maxwell on his desk.

What was particularly remarkable in Maxwell's development of his theory was that it also demonstrated the importance of conceptual *models*, in the development of new theory, in this case the postulated elastic Aether (an invisible medium filling the universe), from which Maxell obtained the equations which predict Faraday's field lines, the forces on charges in complete vacuum! Maxwell was driven to this partly because it was the accepted theory of the day, and partly by the increasing success of the wave theory. Waves in his time were only known to exist in elastic media or fluids. Once Faraday had suggested the key insight that the field lines might vibrate, the way was open to accepting a finite speed of light, and expanding the metaphorical use of the Aether concept, whose scaffolding Maxwell was to steadily discard in later papers. Einstein himself said "imagination is more important than knowledge," knowledge meaning a mere accumulation of facts or experimental evidence. It was said that the personal relationship between Maxwell and Faraday was more cordial than warm, probably on account of Faraday's greater age, but their free communication and respect for each other's work, so evident in their correspondence, was the key to success.

Maxwell's prediction for the speed of light had much to do with the choice of units previously adopted in the independent fields of electrostatics and magnetics—these had to be reconciled with each other when the two subjects merged in the early nineteenth century. It had already been noted before Maxwell began his work that the ratio of the size of a charge expressed in the electrostatic system of units, to the size of the same charge expressed in the electromagnetic system, is a constant with the dimensions of speed, and this constant is approximately equal to the speed of light. This was believed to be a coincidence.

The electromagnetic system (emu) used a definition of amount of charge based on the forces between flowing currents (Ampere's law), while the electrostatic system (esu) was based on the force between static charges (Coulomb's law). Today's unit of charge (the Coulomb), is defined as the amount of charge transported by a current of one amp in one second. (A single heating element on an electric stove draws a current of about twenty amps, or twenty Coulombs per second.) The amount of charge on one electron is about a millionth of a millionth of a millionth of a Coulomb.

Wilhelm Weber (1804–91), after whom the present-day unit of magnetic flux is named, worked with the great mathematician Karl Gauss (1777–1855) at Goettingen University on this problem of the accurate quantification of charge and current. He designed precision instruments such as the galvanometer, an instrument for measuring current like an ammeter. But in 1837, when King Wilhelm IV revoked the liberal constitution of his state of Hanover, Weber signed a letter of protest with other professors, as a result of which they were fired. Weber moved to take up the Chair of Physics at Leipzig, working with Robert Kohlrausch. There, in order to understand the connection between forces between stationary charges (electrostatics) and those between moving charges or currents (electrodynamics), they developed the concept of a current balance, in which the aim was to compare the forces between static charges and flowing currents, in order to reconcile Ampere's and Coulomb's laws. Capacitors (in those days Leyden jars) could be used to store charge, while the new battery provided a source of current. They compared the forces by allowing a measured amount of charge (in electrostatic units, esu) from a Leyden jar to flow as a current, which they measured (in electromagnetic units, emu). Their ratio was approximately equal to the speed of light, however, this was not the focus of their investigations. They did speculate in an 1846 paper, citing Faraday's work on his field lines, that a "neutral electric medium" (the Aether) may support the propagation of light, presumably along vibrations in its tensioned field lines. They had guessed, but had not shown, one of the most important discoveries of nineteenth-century science, that light was an electromagnetic wave. Yet they did not connect this idea with a finite velocity for the transmission of light and, as Filonovich indicates, probably believed light transmission to be instantaneous.

James Clerk Maxwell was born in Edinburgh in 1831 into a prosperous family in which his father, a lawyer, strongly encouraged the boy's interest in engineering and science. His only sibling, a sister, died in infancy. As a child he showed insatiable curiosity for how things worked, as his mother wrote in a letter when he was three years old:

> He is a very happy man, and has improved much since the weather got moderate; he has great work with doors, locks, keys, etc., and "show me how it doos" is never out of his mouth. He also investigates the hidden course of streams and bell-wires, the way the water gets from the pond through the wall ...

As a child he learnt to recite by heart long passages of scripture and poetry, and retained this very detailed knowledge of the Christian scriptures throughout his life. His mother died when he was eight years old, and he had a difficult time at the Edinburgh Academy, being considered backward because of his country Galloway accent and the homemade shoes and tunic he wore on his first day. His nickname "Daftie" followed, but exclusion was relieved by the close friendships he made with Lewis Campbell and Peter Tait, both of whom became scholars and friends for life. Campbell was to write his biography in 1882. Maxwell wrote poetry from an early age, winning the school's mathematics, english and poetry prizes at age thirteen, and his first scientific publication at the age of fourteen on the geometry of ellipses. He was devoted to Scottish poetry, and continued writing poetry throughout his life, perhaps inspired by Robert Burns, sometimes accompanying himself on guitar. His poetry was later published by Campbell.

From 1847 he attended the University of Edinburgh, but also started his own experiments at the family home at Glenlair, resulting in two papers in *Transactions of the Royal Society of Edinburgh*, one of them on the theory of elasticity, which would later be needed to understand waves propagating in the Aether. In 1850 Maxwell moved to Cambridge, firstly in Peterhouse, and then Trinity College, where he was elected to the elite secret society known as the Cambridge Apostles. During his time at Trinity the Great Exhibition at the Crystal Palace in London would have occurred, the first World's Fair. He graduated from Trinity in 1854 with a degree in mathematics, winning the Smith's prize and becoming a Fellow of Trinity the next year. In 1856 he took up a teaching position at Aberdeen, then moved to King's college in London in 1860.

During his time at Marischal college in Aberdeen, Maxwell worked on the problem of the composition of Saturn's rings—what are they? He described them as "a flight of brick-bats." If they consisted of rocks, why didn't they fly away into space or crash into Saturn? His highly mathematical analysis of this difficult *many-body* problem established that the stable arrangement observed was possible. He worked on the problem for two years, a useful preliminary to his later work on the foundations of statistical mechanics. The resulting essay won the Adams prize from St John's College, Cambridge and high praise from the Astronomer Royal, George Airey, himself a leading mathematician. Airey commented that it was

> one of the most remarkable applications of mathematics to physics that I have ever seen.

Only recently have spacecraft visiting Saturn confirmed Maxwell's theory, as shown in the frontispiece. In 1865 he resigned his chair at King's and returned to working alone at Glenlair, until 1871 when he returned to Cambridge to set up and teach at the new Cavendish laboratory. He was closely involved in all aspects of the construction of the new building (now the New Museums site on Free School Lane), and in the selection of apparatus and the curriculum of the courses in physics. Since that time, twenty-nine Cavendish laboratory researchers have won Nobel prizes, and many important discoveries have been made in the building Maxwell designed. These include the discovery of the electron by J.J. Thomson in 1897 and the neutron by Chadwick in 1932, the splitting of the atom by Cockcroft and Walton in 1932, Hodgkin and Huxley's discovery of how nerves propagate electrical and chemical signals, and Watson and Crick's discovery of the structure of DNA.

After devoting much of his time to editing volumes of Henry Cavendish's work, Maxwell died in Cambridge in 1879 at the age of forty-eight from painful abdominal cancer. Those who knew him as a slight figure with the mannerisms of rural Scotland commented on his sharp intellect, social awkwardness, but witty and ironic conversation. His scientific legacy extends far beyond his theoretical work on electro-dynamics and the speed of light, including theory and experiments on color vision, the structure of the rings of Saturn, the kinetic theory of gases and thermodynamics, and one of the first papers on control theory. He was assisted by his wife Katherine in London during the time they conducted the famous experiment showing the counterintuitive result that the viscosity of a gas is independent of pressure. His original apparatus for this experiment, using spinning disks in an evacuated chamber, is now on display at the new Cavendish Laboratory museum in Cambridge, UK. **Figure 6.3** shows James and Katherine in 1869, with their dog Toby. Here's how Maxwell's biographer Campbell describes their life in Scotland:

> *At Glenlair, after Kings, his favorite exercise — as that in which his wife could most readily share — was riding, in which he showed great skill. A neighbor remembers him in 1874, on his new black horse, "Dizzy," which had been the despair of previous owners, "riding the ring," for the amusement of the children of Kilquhanity, throwing up his whip and catching it, leaping over bars, etc. A considerable portion of the evening would often be devoted to Chaucer, Spenser, Milton, or a play by Shakespeare, which he would read aloud to Mrs. Maxwell. On Sundays, after returning from the kirk, he would bury himself in the works of the old divines. For in theology, as in literature, his sympathies went largely with the past. He had kindly relations with his neighbors and with their children and used occasionally to visit any sick person in the village, and read and pray with them in cases where such ministrations were welcome. One who visited at Glenlair between 1865 and 1869 was particularly struck with the manner in which the daily prayers were conducted by the master of the household. The prayer, which seemed extempore, was most impressive and full of meaning.*

Soon after Weber and Kohlrausch's paper, after teaching in Aberdeen, and discussions and correspondence with Faraday, who was about forty years his senior, Maxwell devoted himself to his grand synthesis of the theory of electric and magnetic force fields in a series of three great papers. The first, in 1855 ("On Faraday's lines of force") represents the Aether by tubes of massless fluid, providing an analogy for electrostatics whose properties can then be described by the familiar equations of fluid dynamics. All three papers were all struggling with the question, *what is electricity, and what is the best metaphor to describe it?* Maxwell's second

Figure 6.3 James and Katherine Maxwell in 1869 (with their Scots accents, firm religious convictions, social awkwardness, and ironic wit) and their dog Toby. (From Wikipedia.)

paper ("On physical lines of force") in 1861 adds mass to the fluid requiring Newton's equations to describe it, and a much more complex mechanical model involving vortices, wheels, and tiny gears to fill the vacuum, as will be discussed further. His third paper in 1865 ("A dynamical theory of the electromagnetic field") does away with mechanical models altogether, focusing more on energy transport, the Lagrangian method, induction, and the time dependence of the coupled fields, ending with his prediction for the speed of light.

Maxwell's final famous equations are written out as twenty equations in component form. The use of modern vector notation was just being worked out in Maxwell's time. Using vectors his equations were reduced by Heaviside in 1884 to the four that we teach to students in all electrical engineering and physics courses today, as nicely demonstrated in

T.K Simpson's book. These equations, which can be further reduced to two in relativistic form, provide the basis for all modern electrical engineering, including the design of electric motors, generators, computer data storage, radio communication and antenna design, X-ray science, optics, and much more. Maxwell did suggest names for the required vector operations, including "curl," based on a Scottish game. His third paper does contain one sign error, which has been analyzed in detail for its significance. Taken together, the equations give the direction, magnitude, and time dependence of the magnetic and electric fields in terms of the given charges and currents responsible. The complications he had to deal with arise because the magnetic and electric field strengths depend on each other and their time dependence. For example, a time-varying magnetic field generates an electric field, and vice versa, a significant symmetry which provided an important clue to Einstein in formulating his special theory of relativity.

Maxwell was led to this investigation by the influence of his friend William Thomson (later Lord Kelvin), seven years his senior, who had been the first to try to express Faraday's concept of force fields in mathematical terms in the 1840s. Thomson had found that problems in electrostatics could be solved immediately by analogy with problems in heat flow, for which many solutions had already been given by French mathematicians. Thomson's later (1856) paper on Faraday's magneto-optical effect required the magnetic field to be filled with spinning magnetic vortices which passed their rotational effect onto the passing light waves. The magneto-optical effect (*Faraday rotation*) was the first experimental evidence that light and electromagnetism are related.

Maxwell incorporated these vortices (which he called "idle wheels") into his mechanical model of the Aether, as shown in **Figure 6.4**, taken from his 1861 paper, just after he moved to King's College. The figure shows his lattice of elastic vortex cells surrounded by idle wheels. Lateral motion of the idle wheels represented a current. Strains, due to an electric field, would shift the idle wheels and so produce a momentary current—this turned out to be the entirely new displacement current which was one of Maxwell's most important discoveries. This current explains, for example, how current can flow across the gap between the plates of a capacitor if the voltage on the capacitor changes—a changing field induces a current. It is also crucial to the radiation field we will describe in Chapter 9 on radio waves—it would not be possible to derive the wave equation for electromagnetic waves without this term. The

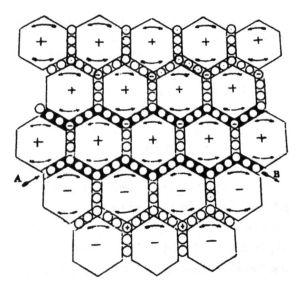

Figure 6.4 Maxwell's model of the elastic Aether with spinning vortices, used for his theory of light waves in his 1862 paper. (From Simpson (2006).)

detailed history of how he discovered this term, and our modern view of it, is given in the article by Bork (1963) listed in the References. The importance of the displacement current, which made Maxwell's equations symmetrical, was not understood during Maxwell's lifetime, and was overlooked in reviews of his work until Heaviside stressed its importance in his book published in 1893. There is no evidence that a desire to make his equations symmetrical and so more elegant was a factor in his introduction of the displacement current—Heaviside was the first to stress and use this symmetry.

Maxwell assumed his Aether was subject to the forces generated by electric and magnetic fields, and used his knowledge of elasticity theory and fluid mechanics to find its properties, and the conditions for transverse waves to be supported by this medium. Maxwell was never considering electromagnetic fields in vacuum, since in his day the Aether was supposed to fill all space, however his third great paper on his equations makes no use of an Aether and is much more formally mathematical. Unable to observe electricity directly, he was using the elastic Aether as a metaphor on which to base a mathematical model.

A much simpler and better mechanical model consistent with Maxwell's equations was later produced by FitzGerald, consisting of an array of wheels connected by rubber bands. Oliver Lodge made many demonstration models of these for teaching purposes. Darrigol's (2012) book on the history of optics contains an excellent review of all the elaborate nineteenth-century French, British, and German mathematical models of the Aether as an elastic medium with exceptional properties, such as support for vortices and the absence of longitudinal modes.

After reading Maxwell's 1861 paper, Faraday wrote to him, asking him to explain the physical meaning of his equations, because

> I have always found that you always convey to me a perfectly clear idea of your conclusions . . . so clear in character that I can work from them.

Maxwell and Faraday were able to meet occasionally—apparently they dined after Maxwell's 1861 lecture at the Royal Institution where Faraday worked. A dense crowd was attempting to leave the building when Faraday, who was well aware of Maxwell's publications on the kinetic theory of crowded gas molecules buzzing around in a bottle, called out to him, "Ho, Maxwell − cannot you get out ? If any man can find his way through a crowd it should be you."

The next year, Maxwell was able to calculate the speed at which these waves would propagate through the Aether, based on these equations, finding that it was equal to the reciprocal of the square root of two known constants, the permittivity of free space and the permeability of free space. These two numbers could be measured using electrical circuitry and are the constants appearing in Coulomb's law and Ampere's law, respectively. This simple result also suggested that *the speed of light did not depend on the velocity of the source of light, or that of a light detector.* That finding established one of the most important paradoxes in nineteenth-century physics, since it was believed that, like waves on a flowing river or light from the headlights of a car, the speed of light should equal the speed of the car plus the speed of light from a stationary car (as described by a Galilean transformation). This was the paradox that Lorentz, FitzGerald, Poincaré, Einstein, and others were to struggle with around the turn of the century.

Maxwell used the best values for the two electrical constants available and found the result approximately equal to the best estimates from

Fizeau, Foucault, and those from the astronomers for the speed of light. His excited conclusion, as he wrote, was

> *We can scarcely avoid the conclusion that light consists in the transverse undulations of the same medium which is the cause of electric and magnetic phenomena.*

This was written in 1862. He had the opposite kind of luck from Newton, who, when he first tested his theory of gravity, used it to estimate the distance to the Moon. Because he unluckily used a very poor estimate for the diameter of the Earth, his Moon distance was widely off, causing him to drop his theory of gravity for twenty years. Maxwell had the opposite experience in 1861, when he wrote to Faraday with his first determination of the speed of light—193,088 miles per second, in almost perfect agreement with the published value of 193,118 miles per second from Fizeau. In fact, both these values were slightly off (the correct value being 186,282 miles per second), but the agreement gave him confidence to continue this line of investigation, unlike Newton. By 1864 he had fully developed and published his *Dynamic theory of the electromagnetic field*. There he writes

> *This wave consists entirely of magnetic disturbances, the direction of magnetization being in the plane of the wave. No magnetic disturbance whose direction of magnetization is not in the plane of the wave can be propagated as a plane wave at all. Hence magnetic disturbances propagated through the electromagnetic field agree with light in this, that the disturbance at any point is transverse to the direction of propagation, and such waves may have all the properties of polarized light. The only medium in which experiments have been made to determine the value of K (a quantity proportional to the squared velocity of electromagnetic waves) is air. By the electromagnetic experiments, Weber and Kohlrausch, v = 310,740,000 meters per second is the number of electrostatic units in one electromagnetic unit of electricity, and this, according to our result, should be equal to the velocity of light or vacuum. The velocity of light in air, by M. Fizeau's experiments, is v = 314,858,000, according to the more accurate experiments of M. Foucault it is v = 298,000,000. The velocity of light in the space surrounding the earth, deduced from the coefficient of aberration and the received value of the radius of the earth's orbit, is v = 308,000,000. Hence the velocity of light deduced from experiments agrees sufficiently well with the value of v deduced from the only set of experiments we as yet possess. The value of v was determined by measuring the electromotive force with which a condenser of known capacity was charged, and then discharging the condenser through a galvanometer, so as to measure the quantity of electricity in it in electromagnetic measure. The only use of light used in the experiment was to see the instrument. The value of v found by M. Foucault was obtained by determining the angle through which a revolving mirror turned, while the light reflected from it went and returned along a measured course. No use whatever was made of electricity or magnetism.*

The agreement obtained seems to show that light and magnetism are affectations of the same substance, and that light is an electromagnetic disturbance propagated through the field according to electrodynamic laws.

Maxwell's theory of electricity and magnetism had predicted that the Aether would support electromagnetic waves which travel at the speed of light, as measured by astronomers, Fizeau and Foucault. This powerfully suggested that these electromagnetic waves were indeed light itself. The theory would later be used for waves at other frequencies, such as a radio waves. Notice his point that "The only use of light (in experiments to measure the two electrical constants) was to see the instruments," while in Foucault's mechanical experiments to measure the speed of light using rotating mirrors, "no use whatever was made of electricity and magnetism." But the ideas in these papers had little impact among scientists, until he put them all together in his great book of 1873, which is still in print.

There is a delightful story, taken from his letters, of his reaction to this discovery, which sunk into him during a period of intense work at Glenlair in Scotland. Having derived his theoretical value for the speed of light, he was excited to compare it with measured experimental values, but those numbers were on documents in his London apartment. There was no working telephone in 1864. So he had to wait, to the end of the summer in great anticipation before catching the steam train for the 400 mile trip to London to find out if his theory was correct.

The two discoveries from these papers—the displacement current and the connection between electrical constants and the speed of light, are perhaps the greatest of Maxwell's contributions to science. In particular, we can trace the discovery of the displacement current, which produced a completely new term in his four equations, to Faraday's discovery of the magneto-optical effect and William Thomson's theoretical treatment of it, which introduced the moving vortices into Maxwell's model for the Aether.

As we have seen, Maxwell was both a great mathematician and also experimental physicist. He therefore determined to measure the two electrical constants, in order to obtain an improved estimate of the speed of light. This he published in 1868, partly as a result of a request from the British Association for the Advancement of Science for a unified system of units for electricity and magnetism. **Figure 6.5** shows the circuit diagram from his 1868 paper, in which he invented a torsion

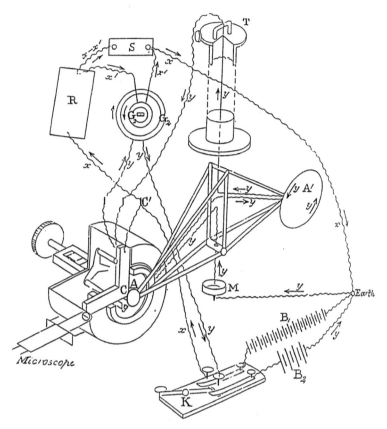

Figure 6.5 Maxwell's balance for measuring the ratio of units in magnetic and electrostatic systems, giving the speed of light and the standard Ohm, as in his 1868 paper. From Simpson (2006).

balance to compare the forces due to static charges with those due to the magnetic field generated by flowing charges in a current. The apparatus cleverly compensates against the Earth's magnetic field, and consists of a torsion pendulum (as in Harrison's chronometers) whose arms balance the forces between static charges on discs A and C (forming a capacitor, charged by a huge battery to several thousand volts!) on one side, and the force between currents in coils, also on the same side. The disk A' on the other side is a dummy needed for gravitational balance. At balance, the electrostatic forces were equal and opposite to the forces between the stationary coil and the moving coil at A. The charge could be calculated from the product of capacitance and voltage. The measured current, voltage, and capacitance (and hence charge) gave values of the two electrical constants mentioned previously. (Maxwell's analysis in his book is far more complex.) After isolating several sources of error, he obtained a value of 2.88×10^8 m/s for the velocity of light, compared with Foucault's 1862 value of 2.98×10^8 m/s, based on the rotating mirror method. The best value from measurements of the aberration of starlight, Bradley's method, gave 3.08×10^8 m/s, but this depended on the accuracy of measurements of the Earth's orbit around the Sun, and its speed. The modern value is 2.99×10^8 m/s, as will be discussed in more detail later.

Maxwell's biographer, Lewis Campbell, gives a charming portrait of Maxwell at work in his laboratory, first quoting from one of his letters in 1878 soon after the telephone was invented:

We have all been conversing on the telephone. Garnett actually recognized the voice of a man who called by chance! But the phonograph will preserve for posterity the voices of our best singers and speakers. I have been making a clay model of Prof W. Gibbs's thermodynamic surface.

Campbell, who knew Maxwell well, writes of him,

He had a strong sense of humor, and a keen relish for witty or jocose repartee . . . his mirth was never boisterous, the outward sign being a peculiar twinkle and brightness of the eyes. Of serenely placid temper, genial and temperate in his enjoyments, infinitely patient, he at all times opposed a solid calm of nature to the vicissitudes of life (such as his painfully protracted death of bowel cancer) . . . In experimental work he was very neat-handed. When working, he had a habit of whistling softly a sort of running accompaniment to his inward thoughts. He could pursue his studies under distractions such as loud conversations. Then he would take his dog into his confidence, and would say softly, at intervals "Tobi, Tobi . . . it must be so. Plato, thou reasonest well". He would then join in the conversation.

The clay model he refers to is now in the museum at the Cavendish laboratory in Cambridge, UK, in addition to the first color photograph, which he invented. We learn that he experienced both the early telephone and the wax cylinder before he died. Around this time, the mayor of Philadelphia is reported to have said that he foresaw a day "when every city in America will have a telephone" (!).

Maxwell also wrote a lot of poetry. At that time, electrical currents were measured by the sensitive mirror galvanometer developed by Kelvin, which used a reflected beam of light instead of a needle to show the current. The light beam would run across the wall of a darkened room as the current changed. Inspired by Tennyson's poem *The Splendor Falls*, he wrote

> *The lamp-light falls on blackend walls,*
> *And streams through narrow perforations,*
> *The long beam trails through paste-board scales*
> *With slow decaying oscillations*
> *Flow current, flow! set the quick lightspot flying,*
> *Flow current, answer light-spot, flashing, quivering, dying*

The story of the contribution of Maxwell's equations to the understanding of the speed of light would not be complete without mention of the extraordinary tale of the 2000-mile-long undersea Atlantic telegraph cable, which was confidently expected to send Morse code messages from New York to London at the speed of light. It did not.

Figure 6.6 shows Kelvin on board the early sail-assisted steamship *Agamemnon* during laying of the first cable in 1857, which broke soon after. Reliable cable for laying finally became available, following about

Figure 6.6 Kelvin (on deck at the center of the group) on board the *HMS Agamemnon* (*US Niagara* in distance) during the laying of the first transatlantic telegraph cable in 1857. Morse code was expected to run under the Atlantic at the speed of light—it didn't, battling the engineers. The first message from Queen Victoria to President Buchanan took sixteen hours for ninety-nine words—at a rate of 0.2 bits/sec! (Provided to the author from US Navy archives.)

five breakages, by around 1866. In some cases a grappling hook could be used to pick up the broken cable from the smooth sandy sea floor to enable its repair. In some sense, this historic image captures the true birth of our global Internet, driven and funded by the desire to send stock market prices and news of military importance between nations more rapidly than the twelve-day transatlantic sailing time. The US ship *Niagra* is visible in the distance. The ships, each carrying half the length of cable, met in mid-Atlantic, where the cables were joined, then payed out onto the sea floor on their journies to Ireland and Newfoundland. Later cables were laid using Brunel's steamship *Great Eastern*, and by the end of the century many nations had laid multiple cables across the Atlantic.

But when the first dots and dashes were sent, Kelvin's galvanometer needle was seen to rise very slowly before equally slowly decaying with each pulse. Why? The great man was consulted to analyze the problem, and quickly realized that such a long insulated wire had capacitance, and this capacitor took time to charge up. Kelvin's subsequent cable equation describing the propagation of pulses along a wire became famous, and was later greatly improved upon by Heaviside with his "telegrapher's equation," where he found a balance condition for distortion-less pulse propagation, by adding inductance. The same equations have since been used to describe electrical signals running along neurons in the brain, and for cable TV signal delivery. Understanding this problem, by use of Maxwell's theory, has indeed made speed-of-light cable communication possible.

It is noteworthy that while Maxwell discovered that light was an electromagnetic wave, there is no mention in any of his papers about the possibility of electromagnetic waves at lower frequencies, such as radio waves, which he could easily have predicted. These were discovered by Hertz in 1887, nine years after Maxwell's death.

We may also wonder how the history of science might have been different had Maxwell read Robert Brown's paper of 1827, in which Brown first observed *Brownian Motion*, or the jumpy motion of pollen particles in water, under his microscope. As Professor Archie Howie at the Cavendish Laboratory in Cambridge has pointed out, having developed the velocity distribution for particles in thermal equilibrium (sharing their energy) in 1860, Maxwell could well have developed the theory of Brownian motion, later worked out famously by Einstein in another of his 1905 papers, providing strong evidence for the existence of molecules and atoms.

7

Albert Michelson and
the Aether Wind

The final decades of the nineteenth century saw a period of intense creative turmoil and debate among scientists in the field of optics regarding the nature of the Aether and its effect on the speed of light. This was mainly due to the work of a supremely gifted young American physicist: Albert Michelson, who well and truly threw the cat among the pigeons, although the significance of his work took decades to sink in. I have a copy of a 1928 textbook on optics written by a student of G. FitzGerald in Dublin (Preston's *Theory of Light*, Preston (1928)), which reviews all the great nineteenth-century speed-of-light measurements, and in which the deeply unsettling effect of Michelson and Morley's astonishing 1887 experiment comes across clearly. Few wanted to believe it, and Michelson himself was disappointed, since he'd set out to find the direction in which the Aether wind was blowing due to the Earth's motion through it. This experiment has been described as "the greatest negative result in the history of science," but its significance and importance grew steadily throughout the final years of the century. It was reviewed in a paper by Lorentz, which was read by Einstein. The issue was one referred to previously—Maxwell's equations said the speed of light did not depend on the speed of the source of light. This was rather like saying that the speed of light from car headlights does not depend on the speed of the car. Rather than accept this inconsistency with the Galilean transformation of velocities, many physicists in the second half of the nineteenth century preferred to believe in the existence of a *stationary* Aether throughout the universe, with respect to which the speed of light was always measured. In view of the success of Maxwell's equations, based (at least initially) on this idea, and the constant speed for light which they predicted, presumably measured in this absolute stationary frame, this was a reasonable view.

Lightspeed: The ghostly Aether and the race to measure the speed of light. John C. H. Spence.
© John C. H. Spence 2020. Published in 2020 by Oxford University Press.
DOI: 10.1093/oso/9780198841968.001.0001

Michelson was to show that they were wrong, and that the measured speed of light on earth was the same in whichever direction it was measured, despite the Earth rushing through space at 67,000 miles per hour. The only conclusion could be that the Earth somehow dragged the Aether around with it, and this was Michelson's conclusion—complete, not partial, drag. The problem was that this was in direct contradiction with Bradley's aberration experiment, in which the direction of light from a distant star changes as the Earth reverses its direction around the Sun. This would not happen if the Earth were dragging the Aether around. The situation was summarized in Larmor's book *Aether and Matter*, written in 1900 It was the intellectual effort to resolve this paradox which led eventually to Einstein's theory of relativity in 1905. The path to that theory in those years was filled with the greatest intellectual excitement among physicists, leading to the birth of both relativity and, indirectly, quantum mechanics.

Michelson was born in Poland in 1853, but his Jewish parents moved to California when he was two. He himself remained agnostic throughout his life. After graduating with honors from high school in San Francisco, he passed an examination for a naval academy, but was not immediately admitted, so he travelled, in one of the first trains across the USA, taking with him a letter of recommendation to President Ulysses Grant. Unfortunately, when he arrived at the White House, the President had just filled his quota of ten appointments-at-large, so Michelson returned to the station, intending to return to San Francisco. Just as the train was about to depart, a messenger from the White House called his name. Grant exceeded his quota and appointed him to the Naval Academy in Annapolis, from which he graduated in 1872, then serving on-board ships for a few years. He later took a position as instructor in physics and chemistry at the Academy, marrying Margaret Hemmingway, daughter of a wealthy New York stockbroker and lawyer, with whom he had two sons and a daughter.

After attending a lecture by the visiting British physicist John Tyndall, a keen promotor of physics in the Victorian era, and reading Maxwell's new text, Michelson decided to focus his research on measurement of the speed of light. As a result of his genius for instrumental design in optics and interferometry he quickly became a leader in that field, improving on Foucault's instruments. His achievements came to the attention of the astronomer Simon Newcomb, with whom he worked before leaving for two years' study leave in Europe, with funding support from Alexander Bell and Newcomb. From 1880 he worked in Helmholtz's laboratory in Berlin, and later in Heidelberg and Paris, meeting Cornu and Lord Rayleigh. On his return from Europe he resigned from the navy and took up a position at Case University in Cleveland, Ohio, where he undertook his most famous experiment. He moved in 1889 to Clark University in Worcester Massachusetts, and in 1892 moved to become head of the Physics department at the University of Chicago. From about 1920, most of his experimental work was done at the Mount Wilson Observatory, near Los Angeles. He won the Nobel Prize in 1907, the first American to do so. He won many other awards, including the Copley Medal of the Royal Society and the Gold Medal of the Royal Astronomical Society, he died in Pasadena in 1931.

Michelson published many papers on his new and more accurate measurements of the speed of light under various conditions using interferometry—in this chapter we are not mainly concerned with those, which were important but incremental advances on the work of Foucault and others. Instead of continuing the theme of direct lightspeed measurements, we turn in this chapter to Michelson's more fundamental experiment, which led to questioning of the very existence of the Aether. But before describing his remarkable scientific career, we need to summarize the status of ideas regarding the existence of the Aether and the science of electricity and optics at the end of the nineteenth century.

By the time Michelson began his experiments at Case, Maxwell's equations had been fully accepted because of their increasingly successful predictive power. They described not only light propagation, but also the propagation of radio waves, discovered by Heinrich Hertz in 1887, and were useful to the rapidly growing telegraph and early electronics and radio industry, as pioneered by Marconi, Fleming, Lodge and others around the turn of the century. There was a telegraph line between Washington and Boston as early as 1846, and we have seen the importance of Maxwell's equations for distortionless transmission of an electrical pulse along a coaxial cable.

Maxwell died in 1879 before either Hertz's discovery or Michelson's work. He was undecided about the Aether, as his theory steadily became more abstract and he discarded this mechanical scaffolding. In his article for the ninth edition of the Encyclopedia Britannica, however, he did propose an experiment to detect the existence of the Aether which followed along similar lines to Michelson's later great experiment of 1887.

Maxwell proposed using the motion of the entire solar system around the center of our Milky Way galaxy to measure our motion through a "stationary" Aether, presumably bolted onto the remotest "fixed" stars, or perhaps the center of our galaxy, and so acting as a God-given absolute frame of reference. The existence of galaxies had only recently been established by Herschel's work in Bristol. Our Sun is just one of many stars making up our Milky Way galaxy, which rotate about a central black hole, rather as the Earth orbits the Sun. Here is what Maxwell says about the Aether, writing in 1878:

> The hypothesis of an Aether has been maintained by different speculators for very different reasons. To those who maintained the existence of a plenum as a philosophical principle, nature's abhorrence of a vacuum was a sufficient reason for imagining an all-surrounding Aether, even though every other argument should be against it. To Descartes, who made extension the sole essential property of matter, and matter a necessary condition of extension, the bare existence of bodies apparently at a distance was a proof of the existence of a continuous medium between them.
>
> But besides these high metaphysical necessities for a medium, there were more mundane uses to be fulfilled by Aethers. Aethers were invented for the planets to swim in, to constitute electric atmospheres and magnetic effluvia, to convey sensations from one part of our bodies to another, and so on, till all space had been filled three or four times over with Aethers. It is only when we remember the extensive and mischievous influence on science which hypotheses about Aethers used formerly to exercise, that we can appreciate the horror of Aethers which sober-minded men had during the 18th century, and which, probably as a sort of hereditary prejudice, descended even to the late Mr John Stuart Mill.
>
> The disciples of Newton maintained that in the fact of the mutual gravitation of the heavenly bodies, according to Newton's law, they had a complete quantitative account of their motions; and they endeavored to follow out the path which Newton had opened up by investigating and measuring the attractions and repulsions of electrified and magnetic bodies, and the cohesive forces in the interior of bodies, without attempting to account for these forces.
>
> Newton himself, however, endeavored to account for gravitation by differences of pressure in an Aether, but he did not publish his theory, "because he was not able from experiment and observation to give a satisfactory account of this medium, and the manner of its operation in producing the chief phenomena of nature."
>
> On the other hand, those who imagined Aethers in order to explain phenomena could not specify the nature of the motion of these media, and could not prove that the media, as imagined by them, would produce the effects they were meant to explain. **The only Aether which has survived is that which was invented by Huygens to explain the propagation of light.** The evidence for the existence of the luminiferous Aether has accumulated as additional phenomena of light and other radiations have been discovered; and the properties of this medium, as deduced from the phenomena of light, have been found to be precisely those required to explain electromagnetic phenomena.

Maxwell's proposal in his Encyclopedia Britannartica article was to use Roemer's method first to measure the speed of light traversing Earth's orbit around the Sun in one direction by observing eclipses of a moon of Jupiter, and then to repeat this when the light coming from Jupiter to Earth was travelling in the opposite direction, after Jupiter had gone halfway round its orbit around the Sun. The orbit of Jupiter takes about twelve years, so these measurements needed to be taken six years apart. In the simplest model, the difference in the speed of light with, and against, the stationary Aether would give us twice the speed with which the Earth and the Sun move through the Aether, confirming its existence. In fact, the velocity of the Sun around the center of our Milky Way galaxy is about 250 km/s, (560,000 mph) which would produce a time difference in these measurements of only 1.6 seconds, assuming the Aether were fixed to the black hole at the center of the Milky Way.

For a simple measurement on Earth, all methods use light sent to a mirror a distance L away, and returning. The time for light to make this round trip is $T_1 = 2L/c$, where c is the velocity of light. Now if the apparatus is moving through the Aether at velocity v (produced, for example, by the movement of the Earth parallel to the direction of light travel), it will experience an Aether wind, and this round-trip time, slowed down when going upwind, but sped up when going downwind, will be

$$T_2 = L / (c - v) + L / (c + v) = 2Lc / (c^2 - v^2).$$

Our instrument must be sensitive enough to tell the difference between these times,

$$\Delta T = T_2 - T_1 = (2L / c) (v^2 / c^2).$$

So the fractional accuracy is $\Delta T/T_1 = v^2/c^2 = \beta^2$. With v = 30 km/s for the Earth's orbital speed about the Sun (about 67,000 miles per hour, much greater than the Earth's daily rotational surface speed of about 1000 mph), we have $\Delta T/T_1 = 10^{-8}$. Unlike Maxwell's "first-order" scheme using Jupiter (for which there could be no return trip from a mirror, and $\Delta T/T$ is proportional to v), it became clear that the earthbound measurements were "second order," requiring apparently impossible accuracy in order to detect a time difference proportional to the square of the already very small quantity β. It was exactly this extremely challenging experiment, in the days before invention of the laser, that Albert Michelson decided to attempt, and which made him famous for an experiment which has forever linked his name with that of Albert Einstein, and for which he invented the interferometer that bears his name.

Michelson got the idea for his great experiment in the spring of 1879, when he read a letter from Maxwell to David Todd, a colleague of Michelson's at the Nautical Almanac Office in Washington. Maxwell had written to ask if the data they had on eclipses of Jupiter's moons was accurate enough to test his idea for detecting the motion of the Earth through the Aether. Maxwell himself had only eight months to live, and Albert Einstein was just five days old. In the letter, Maxwell pointed out the difference between his "first-order" effect and the "second-order" effect of all current methods which depended on a return trip of light from a mirror, making measurements much more difficult. On reading Maxwell's letter, it may have occurred to Michelson that perhaps he could split a beam into two beams at right angles, one running along the Earth's motion and the other across it, and then allow them to interfere. The beam running across the Aether wind would return more quickly than the other beam, just as a swimmer will find it takes less time to swim across a flowing river and back than to swim a distance equal to the width of the river directly upstream against the current and back. This method would allow him to take advantage of the great sensitivity of interference methods which he and Fizeau had found in their previous lightspeed measurements, and so detect a very small second-order effect. Thus was born the concept of the Michelson interferometer. His method may have derived partly from the Jamin interferometer he would have seen in Helmholtz's laboratory. The Jamin interferometer also divides a broad beam by amplitude, like a piece of glass coated with a partially transparent film of silver (a half-silvered mirror) in which all the beam's energy is divided into one of two beams, either reflected or transmitted through the mirror. All the energy in the beam is used, unlike the alternative scheme (known as *division of wavefront*) used by Thomas Young. Here a broad beam of light illuminates two pinholes coherently, but only the tiny amount of light falling on the pinholes gets through to contribute to the interference pattern. Division by amplitude collects energy across the entire wavefront leading to a much brighter interference pattern, and this was a big factor in Michelson's subsequent success. Michelson spent eight years getting his interferometer working properly.

We know that waves running downstream on a river travel with the velocity of those waves in still water, plus the velocity of the current. Was this also true for light? This question of the effect of a moving medium on the velocity of light was a most important one, studied by many researchers in the nineteenth century, and indeed the title of Einstein's 1905 paper refers exactly to this central problem.

Experiments were undertaken to measure Bradley's aberration effect using a telescope filled with water (moving with the Earth) by the Astronomer Royal, George Airey, in 1871, but no change was seen due to the water. The most successful theory to account for the velocity of light in a moving medium (which was assumed to be filled with Aether) was due to Fresnel. This brilliant and early theory (which fortuitously gave the right answer) was known as the *Aether partial drag* hypothesis, and was repeatedly tested by experiment up until the 1920s. Young and Arago had analyzed Bradley's aberration effect using wave theory, assuming a stationary Aether through which the Earth moved. Under that assumption, the velocity of light, fixed with respect to the Aether, would be greater when the Earth was moving toward the star and smaller when it moved away. Arago sought this effect, by measuring the refraction angle of light from stars under these two conditions. (Snell's law has told us that this angle depends on the speed of light.) He found no effect, and, extremely puzzled, asked his friend Augustin Fresnel about this, who replied in a now famous letter with his Aether drag hypothesis.

The idea of Fresnel's Aether drag theory, as mentioned in Chapter 5, was that the Earth drags some of the Aether around with it like a fog, which permeates all materials, and which then moves more slowly with respect to the ground, but has higher density than the Aether in outer space. Only the surplus Aether within the body was supposed to move with the body. Using a mass conservation argument for the Aether, Fresnel found (in 1818!) that the velocity of light within a body (such as water) moving at speed v through the Aether would be

$$c' = c/n \pm v[1 - 1/n^2]$$

where n is the refractive index of the medium and the plus sign is used when light and the body move in the same direction; minus if the motion is antiparallel. The second factor is his *Aether drag coefficient*. The Aether is postulated to be entrained and to move with velocity $(1 - 1/n^2)$ v when a body moves with velocity v. Note that this correction, unlike the effect sought by Michelson, is linear in v, not second order. The

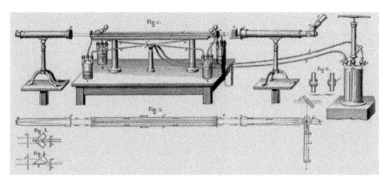

Figure 7.1 Fizeau's 1851 interferometer for measuring the Fresnel drag coefficient. Water flows in opposite directions in the two pipes shown in the middle of "Fig 2." Light from a lamp at the right-hand end is split into two paths through the water streams and reflected back through them again by the mirror at the far left of "Fig 2." This light was then viewed by a microscope off to the side at the right-hand end, through an inclined and partially transparent mirror. (From Frercks (2005).)

famous British physicist, George Stokes, subsequently published the same result obtained by a different method, imagining that the Aether was completely dragged by the Earth, but had a viscosity which depended on the speed of the object, being high at high speed and low at low speed, like pine wood pitch.

For light, a rigid elastic solid was needed for fast transmission in the manner of Bernoulli's *Newton's Cradle*. The result can also be obtained using Maxwell's constitutive equations for light propagation in a solid. It has been commented that Fresnel's analysis results in drag by a medium, not by the Aether, so that an Aether may not be needed. We will see later that, perhaps by an extraordinary coincidence, exactly the same equation is given to an excellent approximation by Einstein's relativity theory in the absence of any Aether. Thomas Young described the Aether around this time (in 1804, before Fresnel's work) as follows:

> *Upon considering the phenomena of the aberration of the stars, I am disposed to believe that the luminiferous Aether pervades the substance of all material bodies with little or no resistance, as freely perhaps as the wind passes through a grove of trees.*

Fizeau showed his gift for interferometry design, anticipating Rayleigh's refractometer, in perhaps his best experiment of all in 1859, as shown in Figure 7.1. This experiment was later repeated by Michelson (using his division of wavefront for stronger fringes) and again by Zeeman in 1914. Light from a common source can follow either one of two paths, before recombining to produce interference fringes. On one path, the light passes through flowing water (filled with Aether) travelling against the current, on the other it travels with the current. The resulting time difference for light waves (either slowed down or sped up by the current) arriving at the detector produces a phase shift, causing the bright and dark interference fringes to move sideways. Fizeau watched these fringes through a microscope as he varied the speed of the water pumped through the system, and found them to move, giving a dependence of the speed of the light on the speed of the water, which contained dragged Aether to support the light waves. He first detected fringe motion with a water speed of 2 m/s, far less than the velocity of light, and an indication of just how sensitive interferometry is to changes in optical path length. (The detection of gravity waves by the LIGO interferometer uses a similar principle.) Notice how his interferometer design, unlike Michelson's later improved method, picks off the light near the source at two points, rather than dividing its amplitude across the whole width of the beam, to give more intensity. If this agreement between experimental results and a theory based on the nonsense of the Aether theories gives you a headache, recall my previous comment about how Fresnel's result is actually fortuitously in agreement with Enstein's special relativity, the correct answer, which we will describe later.

Fizeau's results for his pumped water experiment fitted Fresnel's Aether drag theory within about 15%, but there were difficulties with measurement of the water speed and with its variation across the tubes. But this experiment and Stokes's work did give confidence in the Fresnel partial drag model throughout the second half of the nineteenth century.

More accurate measurements were undertaken by Michelson and Morley in 1886, using a gas lamp, flowing water, and a similar but greatly improved interferometer design, which also supported Fresnel's theory. In 1914, an experiment by Zeeman improved the agreement even further to within 2.6%, some years after Einstein had published his theory, which abolished the Aether altogether!

In spite of all this impressive agreement, these results do not in fact support the Aether drag hypothesis, as was pointed out by von Laue in 1907, soon after Einstein's theory was published. von Laue used Einstein's new theory of special relativity to re-work Fresnel's problem correctly, assuming a relativistic correction, but no Aether. We can anticipate the results of relativity given in chapter 8 here, to show this. If we assume that there is no Aether at all, then the velocity of the light relative to the water is $U = c/n$. If v is the velocity of the water relative to the laboratory, then Einstein's velocity transformation formula gives the velocity of the light relative to the laboratory V as

$$V = [U + v]/G,$$

where $G = 1 + U v / c^2$.

Now if v/c is small (the speed of the water is much less than the speed of light), this can be expanded and approximated by

$$V \sim c/n + v\left(1 - 1/n^2\right),$$

which is the same as Fresnel's expression, assuming Aether drag.

Given this broad acceptance of the Aether drag hypothesis, we can understand the shock in the community when Michelson published his work on the Aether drift, first in Berlin and then with Morley at Case Western University in Cleveland Ohio in 1887, showing that there was essentially no effect on the speed of light due to the Earth's motion through an Aether. The 1887 paper was entitled "The relative motion of the earth and the Luminiferous Aether." Rayleigh, Lorentz, and Kelvin expressed shock and disappointment at the prospect of abandoning Fresnel's theory, and many scientists attempted to find errors in Michelson's work. Michelson and Morley's result also demolished the "stationary Aether" hypothesis (stationary with respect to the distant "fixed" stars), since, in this simplest model, for light travelling in the same direction as the Earth, the full Earth's velocity of 67,000 mph would then have been added to the velocity of the light they measured. The only possibility was "complete drag," meaning that the local Aether at ground level or throughout the universe rotates with the Earth, a highly improbable scenario and one which contradicted Bradley's aberration of star-light effect.

Michelson himself was disappointed, and seemed at first to value the result mainly because it had resulted from his invention of a new kind of interferometer—he was certainly more interested in making accurate measurements of some quantity than getting a null result. He had expected to identify the absolute frame of rest in the universe experimentally from the fringe shifts, with Maxwell's speed of light in this frame adding to the Earth's speed. It is interesting that the Nobel Prize awarded to him in 1907 before Einstein's 1905 theory became widely accepted cites only his precision optical instrumentation "and metrological investigations."

Many scientists attempted to refute Michelson's finding with new experiments. All failed, as summarized at a conference in 1927 at Mount Wilson, California: *The latest Aether drift experiments to date find that the ratio of the velocity of light measured parallel to the earth's velocity and at right angles to it cannot deviate from unity by more than 10^{-15}*. A completely new idea was clearly needed to sort out this mess, which we explore in Chapter 8.

We see here how difficult it was for scientists born before about 1900 to give up on the idea of a fixed reference frame in the universe, in which the speed of light was given by Maxwell's constant value relative to that frame. This was true even for far-sighted scientists such as Lorentz, Michelson, and Poincaré. These highly intelligent scientists were undoubtedly strongly influenced by their experience with fluids, for which Rayleigh's magisterial text and Helmholtz's papers were the authorities. In fluids, the speed of a wave is always measured with respect to the medium in which it is excited, so that currents in a river add to the speed of surface waves, and wind affects the speed of sound relative to the ground. Think of a speedboat cresting smooth waves on a river in which there is a strong current. The same might be expected (incorrectly, in view of Einstein's work) for lightspeed measurements on Planet Earth as it rushes through the fixed reference frame of the Aether. To this should be added the fortuitous agreement of Fresnel's theory with experimental measurements, and Bradley's results on stellar aberration.

Let's summarize what had been learnt by about 1890. Critical to all arguments is the relationship between frequency, wavelength, and speed for a wave. Firstly, there are several separate effects on the speed of a wave when it travels through a medium moving with velocity v with respect to a detector on the ground—the speed of the medium relative to the detector, the speed of the source, and for light, the refractive index of the medium. For light, there were three factors to disentangle—the postulated Fresnel partial drag effect, supported by Fizeau's experiments (which could be anywhere from zero, to Fresnel's suggested value, to complete drag), the aberration of star-light effect showing dependence of lightspeed on detector (Earth) speed, and the Doppler effect.

Now the Doppler effect changes only the frequency of the light, and is caused by relative motion between source and detector. Refraction, by contrast, affects wavelength and speed in such a way that frequency remains constant. Note that in the Michelson experiment, both source and detector are fixed, so there is no Doppler effect. It was understood that the speed of waves on the surface of a river adds or subtracts to that of any underlying current. The speed of sound is a constant relative to the medium in which it is moving—on a windy day, this medium is moving, and the sound waves move with it. But the speed of sound waves is independent of the speed of their source, whereas a moving source will change the frequency via the Doppler effect.

There were many contradictions among scientific papers, in particular, Michelson had interpreted his 1886 improvement on Fizeau's flowing water experiment as supporting Fresnel's stationary Aether (and partial drag) hypothesis, but his famous 1887 experiment proved that a stationary Aether could not exist, and was at first (incorrectly) interpreted in terms of Stoke's complete drag theory. To add further confusion, a very interesting and visionary proposal by Woldemar Voight appeared in the same year, based on a calculation of the Doppler effect for waves in an incompressible elastic Aether, using velocity transformations, which left the wave equation unchanged in the moving medium. His theory, which was largely ignored, therefore contained the correct relativistic notions of length contraction and time dilation and, as we shall see, explained Michelson's 1887 null result.

After his preliminary experiments in Berlin in 1881, Michelson published his results from the interferometer he had designed in Potsdam which was built in Germany, using optical components from Breguet in Paris. He concluded that

> ... the interpretation of these results is that there is no displacement of the interference bands. The result of the hypothesis of a stationary Aether is thus shown to be incorrect.

By "stationary" Michelson is referring to an Aether fixed to the distant stars. His experiment showed that there was no "current" or "wind" due to the Earth's motion through a stationary Aether. The results were noted by Rayleigh, Kelvin, and Lorentz, but Michelson himself was disappointed. It was later found by Lorentz that he had made an error in the analysis and overestimated the sensitivity of the method by a factor of two. As a result, discouraged, he abandoned this line of research on returning to the USA, considering the work to be a failure.

It was his friendship with the older and highly regarded Morley (a chemist, physicist, and mathematician at Case Western University) and Kelvin's visit to Baltimore to lecture at Johns Hopkins University in October 1884, where Kelvin and Rayleigh urged Michelson to try his interferometry again, that convinced Michelson and Morley to try again with a better design. This they did on their return to Cleveland. But first, they repeated more accurately the flowing water experiment of Fizeau, again using their improved division of amplitude method. The result, which he immediately sent in letters to Kelvin and Willard Gibbs, showed that the light speed depended on the water speed in exact agreement with Fresnel's partial Aether drag theory. Around this time, Michelson wrote to Rayleigh:

Cleveland, March 6, 1887.

My dear Lord Rayleigh,

I have never been fully satisfied with the results of my Potsdam experiment, even taking into account the correction which Lorentz points out.

All that may be properly concluded from it is that (supposing the Aether were really stationary) the motion of the earth thro' space cannot be very much greater than its velocity in its orbit.

Lorentz's correction is undoubtedly true. I had an indistinct recollection of mentioning it either to yourself or to Sir W. Thomson when you were in Baltimore.

It was first pointed out in a general way by M.A. Potier of Paris, who was however of the opinion that the correction would entirely annul any difference in the two paths; but I afterwards showed that the effect would be to make it one half the value I assigned, and this he accepted as correct. I have not yet seen Lorentz's paper, and fear I could hardly make it out when it does appear.

I have repeatedly tried to interest my scientific friends in this experiment, without avail, and the reason for my never publishing the correction was (I am ashamed to confess it) that I was discouraged at the slight attention it received, and did not think it worth while.

Your letter has however once more fired my enthusiasm, and it has decided me to begin the work at once.

If it should give a definite negative result then I think your very valuable suggestion concerning a possible influence of the vicinity of a rapidly moving body should be put to the test of experiment; but I too think the result would be negative.

If this is all correct, then it seems to me the only alternative would be to make some experiment at the summit of some considerable height, where the view is unobstructed at least in the direction of the earth's rotation.

The Potsdam experiment was tried in a cellar, so that if there is any foundation for the above reasoning, there could be no possibility of obtaining a positive result.

I should be very glad to have your view on this point.

I shall adopt your suggestion concerning the use of tubes for the arms, and for further improvements shall float the whole arrangement in mercury; and will increase the theoretical displacement by making the arms longer, and doubling or tripling the number of reflections so that the displacement would be at least half a fringe.

I shall look forward with great pleasure to your article on "Wave Theory" (hoping, however, that you will not make it too difficult for me to follow).

I can hardly say yet whether I shall cross the pond next summer. There is a possibility of it, and should I come to your pass I shall certainly do myself the honour of paying you a visit.

Present my kind regards to Lady Rayleigh and tell her how highly complimented I felt that she should remember me.

Hoping soon to be able to renew our pleasant association, and thanking you for your kind and encouraging letter,

I am

Faithfully yours,
Albert A. Michelson.

Figure 7.2 shows the ray paths for Michelson's famous interferometer in simplified form, invented by him and used for this experiment at Case Western University. (In fact he used many mirrors at the sides, so that multiple reflections increased the path length for the light.) **Figure 7.3** shows the actual instrument as built in the basement of Adelbert Hall, and a cross section through the apparatus in its final form. It floats on a ring of liquid mercury (Morley's idea).

We can analyze the simplified version of Michelson's interferometer as follows. Light starts from the source S, half of which is reflected by the beam-splitter B to the mirror M2, the other half to M1. These beams return to bounce off B again and interfere at the screen D, where interference bands are produced because the light is coherent. But the Earth, going at 67,000 mph, moves the splitter B during the time the beam is going to M1 and back, whereas the light is fixed to the stationary Aether filling the universe. By rotating the entire interferometer by ninety degrees about the vertical, the direction from B to M2 could be turned from running along the Earth's motion through the Aether to running across it. The result would be that the Aether wind would move the interference bands sideways by different amounts. This Michelson did not see happen. So in answer to our earlier question—light is *not* like water waves, where the speed of water waves on a river includes the speed of the current. Instead *there is no preferred reference frame for light propagation and no "current."*

In more detail, in the previous discussion comparing Maxwell's proposed use of the moons of Jupiter with "second-order" terrestrial

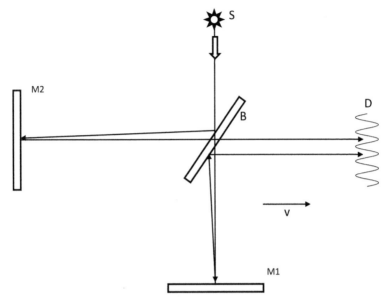

Figure 7.2 Light-ray paths for Michelson's interferometer, simplified, with no Aether effect on the ray paths. Light starts from the source S, half of which is reflected by beam-splitter B (a half-silvered mirror) to mirror M2, the other half to M1. These beams return, to bounce off B again and interfere at the screen D, producing interference bands. These did not move when the entire interferometer was rotated, as they would be expected to if the universe were filled with a stationary Aether in which the Earth moves with velocity v as shown, while light waves remain stuck to the stationary Aether. The Earth's motion changes the light paths when the beam-splitter moves along direction v while light is travelling from B to M1 and back. Einstein concluded that there was no Aether.

mirror experiments, we showed how the time for light to traverse the "upwind" path here from B to M2 in Figure 7.2, and back again downwind is

$$T_2 = 2Lc \, / \left(c^2 - v^2 \right),$$

where v is the speed of the Earth, c the speed of light in the stationary Aether, and L the distance between M2 and B. Using reasoning similar to that used to analyze the aberration of star-light (where the detector on Earth is also moving *across* a beam of light), we find that the time for

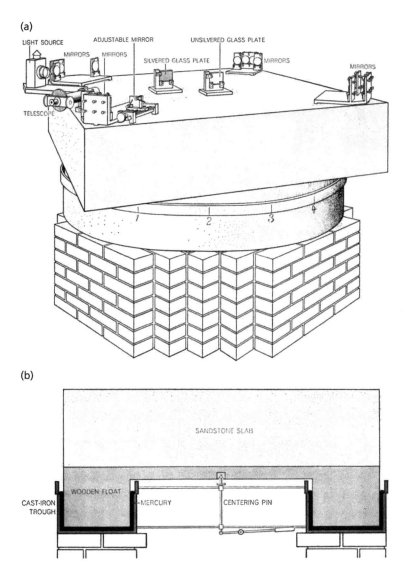

(a)

LIGHT SOURCE   ADJUSTABLE MIRROR   UNSILVERED GLASS PLATE
      MIRRORS   MIRRORS   SILVERED GLASS PLATE   MIRRORS
                                                        MIRRORS
TELESCOPE

(b)

SANDSTONE SLAB

WOODEN FLOAT
CAST-IRON   MERCURY   CENTERING PIN
TROUGH

Figure 7.3 (a) Michelson's interferometer, used to detect any motion of the Earth through the Aether—"the greatest null result in the history of science." To reduce vibration, the optics were mounted on a five-foot-square sandstone slab a foot thick, and the supporting brick pier reached down to bedrock. To allow it to be rotated smoothly, the interferometer floated on a trough of liquid mercury, which also relieved any stresses in the materials. (b) The base of the interferometer, floating in a mercury trough. (From Shankland (1964).)

light to travel across from B to M1 and back (the same time for each trip), is

$$T_1 = (2L/c) / (\sqrt{(1-\beta^2)}),$$

where $\beta = v/c$. Using some approximations, possible because β is small, we find that the phase difference (which needs to change by π to move the interference bands sideways by the width of one band) is

$$\theta = (2\pi/\lambda) L\beta^2.$$

Here, λ is the wavelength of the light, about a thousandth of a millimeter. Making $\theta = \pi$ (or less) and v the velocity of the Earth, Michelson could calculate how long (L) his interferometer needed to be, so that an interference band would move when it was rotated in the Aether wind. He chose L to be about thirty-six feet in total for the Case instrument. In Berlin, using monochromatic sodium lamp light in Helmholtz's laboratory with his first, much smaller instrument, he found the interference fringes disturbed by vibration from horse-drawn traffic outside the laboratory.

After two years of attention to the reduction of vibration, and heating effects which change the length of the arms of the interferometer, the instrument at Case was finally ready to use. The idea was to set it gliding slowly around (about one revolution every six minutes) and read off the fringe positions during rotation over a full year, in order that any motion of the entire solar system through space would cancel out. In July 1887 they recorded their first measurements and obtained the results that they published on July 8, 9, 11, and 12. Nothing happened. The interference bands did not move as the machine rotated—Michelson wrote that *any displacement (of the bands) due to the relative motion of the earth and the luminiferous ether cannot be greater than 0.01 of the distance between the fringes.* They knew they could detect an Aether wind as small as 5 km/s, whereas the Earth's speed is about 30 km/s.

There was some small movement of the fringes as the instrument was rotated, but this was less than 1/20th of that expected if the Aether were truly stationary with respect to the Sun. This was subsequently explained as an experimental artifact due to pressure and temperature fluctuations. By continuing these observations throughout the year as the Earth orbited the Sun and rotated on its axis, they were comparing the speed of light in two orthogonal directions which spanned all possible directions relative to the fixed Aether—at no time did they see an appreciable fringe shift.

The results showed that the speed of light is the same in all directions, and that the static Aether (absolute frame of rest) hypothesis, combined with vector addition of light and medium speeds, was untenable. The experiment does not measure the speed of light, nor support Einstein's later hypothesis that the speed of light is independent of source speed, even if the Earth's speed in orbit changes, and we ignore the accelerations of the Earth's rotation and solar orbit. (This independence of the speed of light on source speed was not demonstrated experimentally until 1964, by Alvager and colleagues) The source and detector are fixed in this experiment.

While Michelson's discovery was a momentous event in the history of science, it was nevertheless a *negative* result. They had been expecting otherwise, and disappointment becomes clear in Michelson's writings. Most scientists were reluctant to accept the result and give up the stationary Aether idea, and at Kelvin's urging, they repeated the experiment in 1904 with much greater accuracy, again finding no shift in the fringes. Lorentz wrote to Rayleigh in August 1892 about the experiment:

> ... I have read your note with much interest, and I gather from it that we agree completely as to the position of the case. Fresnel's hypothesis, taken conjointly with his coefficient $1 - 1/n^2$, would serve admirably to account for all the observed phenomena were it not for the interferometer experiments of Mr. Michelson, which has, as you know, been repeated after I published my remarks on its original form, and which seems to decidedly contradict Fresnel's view. I am totally at a loss to clear away this contradiction, and yet I believe that if we were to abandon Fresnel's theory, we should have no adequate theory at all, the conditions under which Mr. Stokes has imposed on the movement of the Aether being irreconcilable to each other.
>
> Can there be some point in the theory of Mr. Michelson's experiment which has as yet been overlooked?

To further complicate matters, a controversy was also developing at this time as to what was actually being measured. It turned out that there were actually two different light velocities to consider (or, under some conditions, no definable velocity at all). These two velocities are called the group velocity, which is the speed of a pulse of light; and the phase velocity, which is the speed at which the crests move in a long wave. The issue was resolved by Rayleigh with lasting consequences for relativity and communications, and arose from an experiment by Young and Forbes. They used a variant of the toothed wheel method in

1882 to measure light speeds, from which they concluded in their *Nature* paper that blue light travels faster than red light.

Lord Rayleigh (1842–1919), third baron of Rayleigh and, like Maxwell, another winner of the Smith Prize for mathematics, was born William Strutt. He was awarded the Nobel Prize in 1904 for his discovery of the gaseous element argon, and was responsible for many fundamental advances in optics, acoustics, thermodynamics, and the theory of light scattering, for example, explaining the blue color of the sky. In a letter to *Nature* commenting on the Young and Forbes result, Rayleigh suggested that all of the measurements of the speed of light (except for Bradley's stellar aberration) actually measured the speed of a pulse of light (the group velocity V_g), not the speed of the wave crests (the phase velocity V_p). He considered that Bradley's experiment had measured the phase velocity, but the angle Bradley measured required an estimate of the Earth's velocity to convert it to an estimate of the speed of light. Using the available values for the Earth's speed, he found that Bradley's estimate of the phase velocity agreed pretty well with the terrestrial group velocity values. Rayleigh also found a more general relationship between the group and phase velocities. There had also been earlier treatments of this topic published by Stokes and by Hamilton. Rayleigh commented that if a pulse of waves is launched on a still lake, it will be seen that some wavefronts appear to run rapidly through the pulse, then die out. This can be understood if we imagine a pulse to be made up of many sinusoidal waves of different frequencies f (in cycles per second) and hence periods $T = 1/f$. The frequency measured in angular radians per second is $\omega = 2\pi f$. The phase velocity for a single frequency, the speed of the crests, is always the frequency multiplied by the wavelength λ, so $V_p = f \lambda$. It is common to define a *wavevector*, $k = 2\pi/\lambda$, so that $\omega = k V_p$, which is known as the dispersion relation, in this case for light in vacuum. In order to localize a pulse we cannot use a pure sine wave (a wave of single frequency) because a true sine wave must last forever and have no beginning or end (and fill the universe!). It cannot be used to send a signal consisting of the short pulses used for Morse code or binary digits. At around the time of the French Revolution, Joseph Fourier showed that if a piece is cut out of a sine wave, new sine waves with higher frequencies and shorter wavelengths are generated. In other words, in order to localize a short pulse in time, such as Morse code or the transmission of binary digits, we must use a range of frequencies.

If a long rope is secured to a wall at one end and held at the other, a quick flick of the wrist will send a pulse down the rope. The speed of the pulse is the group velocity V_g. Alternatively, if the free end of the rope is smoothly oscillated up and down, continuously, waves of pure sinusoidal shape will be seen to run down the rope. The speed of one crest of such a wave is the phase velocity V_p. This can exceed the speed of light, whereas the group velocity cannot. Rayleigh had shown in 1871 that these are related by

$$V_g = d\omega/dk = V_p - \lambda\, dV_p/d\lambda,$$

where the last term on the right indicates how the phase velocity varies with the wavelength.

In Fizeau's experiment, his toothed wheel clearly chopped up the light into pulses, as did the rotating mirror. By chopping up light, we must necessarily introduce more wavelengths to form a pulse—the light is no longer monochromatic (single frequency), like a single note on the flute. The flute produces perhaps the most pure and monochromatic tones of all the orchestral instruments. The characteristic tone of other instruments is created by mixing many frequencies about the main one.

If the pulses are already travelling in a medium, such as the flowing water of Fizeau or Michelson's experiment, things get very complicated, because the refractive index of the water varies with wavelength, each component frequency travelling with a different speed. The refractive index gives the ratio of phase velocities, not group velocities. This problem of the propagation of polychromatic light in a dispersive medium has been studied extensively, and lies beyond the scope of this book. In extreme cases it can lead to a situation where it is not possible even to define a speed of light.

Rayleigh's critical response to the Young and Forbes paper led to a bitter dispute, which climaxed at the 1884 Montreal meeting of the British Association for the Advancement of Science, under its President, Lord Rayleigh, and which also involved FitzGerald. It had the important outcome of instigating an extended correspondence between Rayleigh and Michelson.

This international meeting, and Rayleigh's discussion of his work, also brought Michelson's work to the attention of many leading physicists for the first time. The meeting stimulated Michelson to measure light speeds (group velocities) in media with large refractive indices, and to compare these with the group velocity given by Rayleigh's formula, using the measured refractive indices to provide phase velocities. Papers from both Rayleigh and the great American physicist Willard Gibbs (1839–1903) immediately followed, analyzing his results. The important conclusion followed that the concept of group velocity could be applied to light, as it had been to other forms of wave motion. Ultimately, this would have to be based on the theory of light scattering in media. This would not be fully developed until the classical theories of Helmholtz, Larmor, Drude, and Lorentz at around the turn of the century, and later developments in quantum mechanics by Sommerfeld and others, most of which depended on the discovery of the electron in 1897.

The group velocity debates and Rayleigh's paper had very important consequences for the possibility of faster-than-light transmission of information, and for certain paradoxes in relativity theory. For it was soon shown that energy and information can only be transmitted at the group velocity (which cannot exceed c), not the phase velocity (which can exceed c if the dispersion is anomalous). These results have profound implications for communications.

To transmit information, we must start and stop the wave (as for Morse code or digital bits) and this immediately introduces a range of frequencies, needed to form a pulse. Normally Rayleigh's formula shows that the group velocity is less than the phase velocity—they are only equal for a sinusoid.

Under rare conditions of anomalous dispersion, when the refractive index increases with wavelength, the group velocity can exceed the phase velocity (but not exceed c), and the phase velocity *can* exceed the velocity of light. We will see that Einstein's discovery that nothing can exceed the speed of light refers to the group velocity (at which information is transmitted). In vacuum, where the dispersion relation $\omega = k V_p$ applies, the frequency of the wave is proportional to the wavevector. Under these conditions the group and phase velocities are equal, and a pulse of any shape travels without distortion. Otherwise, without this proportionality, pulses distort as they propagate.

Recently, several groups have used dispersion effects to produce "slow light," meaning a low speed for the group velocity. Speeds as low as about 38 mph have been obtained (compared with about 186,000 miles per second in free space), and it has even proven to be possible to bring light to a complete halt in a cloud of very cold atoms, then re-start it. The dispersion *constant* gives the variation of the refractive index, $n(\lambda) = c/V_p$, with wavelength for a medium. From the previous equation we can obtain the following:

$$c/V_g = n - \lambda \, dn(\lambda)/d\lambda .$$

The group velocity is less than the phase velocity if the speed of longer wavelengths is greater than that of shorter wavelengths, and this occurs if the dispersion constant $n(\lambda)$ decreases with increasing wavelength, a situation known as anomalous dispersion. Then the group velocity V_g becomes much less than c in this equation, since the last term is large and negative. The underlying reason that it is possible to slow light down when it travels through special materials is that, according to the theory of light scattering from molecules which Rayleigh developed, light is both absorbed and then re-radiated at the same frequency by atoms. This process changes the speed of light by an amount which depends on how close the light frequency is to the resonant frequency of the electrons bound to atoms in the solid. Rayleigh's theory was eventually able to explain the blue color of the sky on this basis.

This explanation for the blue color of the sky makes an interesting story in itself, despite an alternative modern interpretation. John Tyndall was a British experimenter and vigorous promoter of science in the nineteenth century, who toured extensively in Europe and the United States. In 1859 he discovered that, when sunlight light passes through a liquid containing a suspension of particles such as milk, blue light is scattered to the side more than red light. This *Tyndall effect* was later explained by Rayleigh in three papers between 1871 and 1889, providing perhaps the first theory of light scattering by molecules. In these papers he showed that the amount of light scattered is inversely proportional to the fourth power of the wavelength, just as FitzGerald showed for radio-wave emission from a dipole radiator in 1883, as we will see in Chapter 9. Since blue light has shorter wavelength (higher frequency) than red light, more of it is scattered to the side of the molecule. Much debate arose as to the nature of the molecules or dust particles responsible—Einstein's 1911 paper supported the idea that it was nitrogen and oxygen molecules that were responsible. When we look directly above us into the sky when the sun is off to the side, our eyes are detecting light scattered by these molecules as dipole radiation through an angle of about ninety degrees. Then there is about ten times more blue light scattered through this large angle into our eyes than red light. The sky is not violet (even shorter wavelengths) because the sun emits less violet light than blue, and our eyes are less sensitive to violet. But notice that near sunset, when we are looking more directly into the sun, so the scattering angle is small, the sky appears either yellow or red—we are seeing more unscattered red light coming directly towards us, while the blue light is scattered off to the side.

In the final years of the nineteenth century, attempts to explain Michelson's result produced a frenzy of highly imaginative intellectual activity. The conclusion which could be drawn was that the stationary Aether theory was wrong—the speed of light was the same in every direction on the moving Earth. The alternative "complete drag" conclusion that the Aether is fixed to Planet Earth was considered too implausible to take seriously. But it is important to note that, as we have mentioned, in Michelson's experiment, the light source and the detector are fixed, so that the experiment does not show that the speed of light is independent of source or detector speed.

At this point the many theoretical proposals to save the Aether theory produced at the end of the nineteenth century become rather technical and complicated, so you may want to skip the rest of this paragraph. Everyone agreed that light, with its finite velocity, required a medium within which to propagate. In brief summary, Fresnel's theory, with partial Aether drag, accounted for Fizeau's experiment with flowing water and Arago's finding that Bradley's aberration was unaffected by a prism in front of a telescope (changing the light speed); while Stokes's theory, with an Aether completely entrained in and around matter, explained the Michelson and Morley (1887) results. Fresnel's theory accounted for all the failures in detecting Aether drift which were sensitive to first order in velocity, and with stellar aberration, for reasons we will explain later in connection with relativity. However, his theory predicts a positive result with sensitivity to second order effects, which later interferometry experiments did not show. Stokes's theory assumed complete dragging in matter, falling off to zero at large distance. Hertz produced a complete drag theory, based on Galilean transformations in 1890. Stokes's complete drag theory was not supported by the results of the Sagnac interferometer, which sends light in counter-propagating directions around a circle to interfere (and which can be used for navigation). Oliver Lodge constructed an extraordinary apparatus to test Stokes's theory. This placed an interferometer between heavy disks rotating at high speed, and found no influence of the disk speed on the interference. Gustaf Hammer conducted a related experiment in 1935, moving large masses up to an interferometer to no effect. And complete drag is inconsistent with Bradley's aberration, since it would mean that the telescope in **Figure 4.2** would not need to be tilted if the Aether, supporting light propagation, moved with the Earth. Stokes modified his theory in 1845 to account for aberration, by making his Aether both incompressible and irrotational, but a paper by Lorentz in 1886 found that such an Aether could not possess both the required tangential and normal velocity components provided by the Earth's motion. Additional theories were produced by Wilhelm Wein and Theodor des Coudres in (1900) and by Max Planck in 1899, who modified Stokes's theory in such a way as to agree with most experimental results known at that time. This Stokes–Plank theory was not fully refuted by experiments until those of Michelson and colleagues in 1925. Lorentz himself produced a stationary Aether theory in 1892, which was inconsistent with the Michelson–Morley result. He therefore adopted the length-contraction hypothesis, independently of FitzGerald, and combined this with his Lorentz transformations, but retained an Aether.

For a more human story, we must now turn to the insufficiently appreciated achievements of the eccentric and reclusive "amateur" British scientist Oliver Heaviside (1850–1925). Heaviside reduced Maxwell's twenty equations to the modern set of four by co-developing vector calculus. In 1900 he wrote the first widely used textbook on electrodynamics, providing crucial background to Einstein's theory with its astonishing proposal that things get shorter as they go faster, an idea completely unacceptable to the general public and to most scientists at the time.

Heaviside was self-taught and no respector of authority. He started out as a telegraph boy and electrician supported by Wheatstone, then spent years studying Maxwell's book and publishing his developments from it in a popular trade magazine *The Electrician*. The editor supported him for regular contributions of his highly mathematical material by paying him just enough to live on in poverty and poor health. He never married.

Heaviside feuded for years with the electrical engineer William Preece of the British Post Office, which controlled telegraphy and consistently opposed his ideas and career advancement. Matters came to a head in 1887 when Preece blocked his paper proposing inductive loading on transmission lines, a highly successful idea, strongly opposed by Preece, which later made fortunes for the American telegraph companies by reducing the distortion in telephone conversations and increasing the speed of transmission across the Atlantic cable tenfold. Inductive loading is now considered the most important development in telegraphy between Bell's invention of the phone and the first vacuum tubes in 1912.

Heaviside's papers in *The Electrician* were largely ignored by academics and were eventually terminated, probably due to Preece's influence on the new editor. In spite of this opposition, Heaviside's work gradually came to the attention of some eminent figures, including Kelvin, who for a long time was skeptical towards Maxwell's equations, Lodge, and FitzGerald in Dublin, who all supported publication of his work, which started to appear in leading scientific journals from about 1888, gaining him ever increasing recognition. Heaviside is famous for proposing the Heaviside layer, which reflects radio waves from the upper atmosphere and so explains why radio communication is possible beyond the horizon. His equation for the design of electrical cables which avoid distortion is more important than ever today for wi-fi modem and cable TV connections. He invented new mathematical functions, and independently discovered the Poynting vector, which describes the energy flow in electric fields. He was the first to emphasise the symmetry in Maxwell's equations. This results from Maxwell's addition of the displacement current term, and is responsible for alternating current flow across the insulating dielectric slab in a capacitor. "The Aether is dielectric" said Heaviside. He became a Fellow of the Royal Society in 1891.

In 1888, Heaviside showed how the electric and magnetic fields around a charge become deformed as they move through a medium. His 1889 paper in *Philosophical Magazine* also predicted radiation from a moving charge, which occurs if the speed of the particle exceeds the phase velocity c/n of light in the medium. This was later discovered independently by Cherenkov and Vavilov in 1934. The normally spherical surfaces of constant electric field around a stationary charge bunch up into an ellipse, with short axis in the direction of movement, as it speeds up. Remarkably, Heaviside's equation for this velocity-dependent field is exactly that given by Einstein's later theory of relativity, including what later came to be known as the FitzGerald–Lorentz contraction, describing how things get shorter as they go faster. This contraction had been obtained by Heaviside directly from Maxwell's equations, showing that they somehow incorporate relativistic effects. Heaviside sent his paper to Kelvin, H. Hertz, George FitzGerald, and also J.J. Thomson, who later discovered the electron. Correspondence with FitzGerald developed, and they eventually met in London in 1889, a year after Hertz's paper announcing the discovery of radio waves (described by Maxwell's equations) and two years after publication of the puzzling Michelson–Morley result. At that time FitzGerald considered this result to be an "outstanding problem" confronting Maxwell's theory.

In Spring 1889, FitzGerald visited his good friend Oliver Lodge in Liverpool, at 21 Waverley Road, to discuss the Michelson–Morley result. At that time, the idea that matter consists of atoms, separated by electrical fields and forces, was not firmly established. Indeed, Kelvin opposed it well into the twentieth century, until the discovery of X-ray diffraction from crystals. But it occurred to FitzGerald at that meeting with Lodge that, if Michelson's interferometer were moving at the Earth's speed of 67,000 mph through the Aether, then the fields around all of its atoms would contract, bringing the atoms closer together, according to Heaviside's calculation, and making the interferometer itself shorter in the direction of motion. The amount of this contraction, $\sqrt{(1 - \beta^2)}$, was exactly the amount needed to cancel the effect of the Aether drag, explaining the null result of Michelson and Morley. One simply needed to assume that the forces between atoms in all matter are electromagnetic in origin. Hendrik Lorentz, at Leiden in 1892, had independently arrived at the same conclusion.

George FitzGerald (1851–1901) is described by Bruce Hunt as "the soul of the Maxwellian group" of scientists, including Lodge, FitzGerald, Heaviside, Larmor, and Hertz, who developed Maxwell's legacy of ideas until the end of the century. They thereby founded, with Lorentz, the modern field of *classical electrodynamics*. FitzGerald was born in 1851 into a family among the intellectual elite of Dublin; his father was a Bishop and Professor of Moral Philosophy and he quickly excelled after entering Trinity College at the age of 16. Eventually, after a demanding four-day examination in the physical sciences and mathematics, he was awarded a college fellowship in 1877, providing him with lifetime tenure and a solid income, becoming Professor of Natural Philosophy until his death in 1901. His mother, from whom he is thought to have inherited his mathematical abilities, died when he was eight—her brother

George Stoney FRS was responsible for naming the electron. He was an excellent athlete with a strong sense of humor. Heaviside said of him that he had

the quickest and most original brain of anybody, however he saw too many openings. His brain was too fertile and inventive.

Generous with his ideas, he comes across, with Lorentz, as perhaps the most attractive personality in our story—a highly creative individual full of ideas, and more concerned with stimulating ideas among others to advance the field than self-promotion. Lodge said of him that he was "the life and soul of debate," always willing to give up his time for discussion and comment on other people's work at the British Association for the Advancement of Science meetings.

FitzGerald's papers were, as Heaviside commented "not large in bulk, but very choice and original." They included his paper, famously refereed by Maxwell in bed just before his death, on the treatment of reflection and refraction at interfaces using Maxwell's equations. He also wrote important papers on the theory of the Faraday effect (the rotation of the plane of polarization of light by a crystal in a magnetic field) and the Kerr effect (a change in refractive index due to an applied electric field). He made a crucial contribution to Larmor's theory of the electron, and developed the theory of electromagnetic radio wave generation prior to Hertz's discovery of radio transmission in 1887, as we shall see in Chapter 8.

In one of the most overlooked papers in the history of science, FitzGerald unwisely sent his paper predicting length contraction at speed to the American journal *Science*, perhaps hoping that Michelson would see it. At that time *Science* was certainly not the prestigious journal it is today, and the paper was completely ignored. Even FitzGerald forgot he had written it, and it was omitted from his collected writings. The paper is not mentioned in my copy of his student Preston's book on the theory of light, which was written soon after, and indeed FitzGerald never saw his own paper. But Oliver Lodge chose to mention it in his *Nature* paper of 1892, the first appearance of the length contraction hypothesis in print in Europe, one of the most surprising and widely opposed ideas in Einstein's later theory, where the same effect arises for somewhat different reasons.

Before discussing Einstein's theory and its relationship to the speed of light, we should note that another of its most counterintuitive predictions had also been suggested around this time—that of time dilation. Time slows down, not only at traffic lights and boring meetings, but also if you go really fast, relative to the people at home. The great Dutch physicist Hendrik Lorentz (1853–1928) had proposed both the length contraction idea and the dilation of time in 1899, and these ideas were also pursued by the famous French mathematician Henri Poincaré. Poincaré, who always believed in a form of Aether, coined the term *postulate of relativity* before Einstein, and had also written that "there must arise an entirely new kind of dynamics, which will be characterized above all by the rule that no velocity can exceed that of light," and, in 1904, "as demanded by the relativity principle, the observer cannot know whether he is at rest or in uniform motion." Einstein worked as a patent attorney in 1905 and had difficulty accessing a good academic library during working hours, and was probably unaware of Poincaré's work, which anticipated much of the subsequent mathematical development of special relativity. Darrigol, in his excellent book, indicates that Poincaré, while always retaining his belief in the Aether as an absolute frame of reference, went so far as to suggest that, although it existed, its existence could never be detected by any experiment—a view very close to Einstein's founding basis for relativity.

Time dilation had also been suggested independently by Larmor, whose highly abstract papers many found difficult to understand. They all suggested that time passes more slowly for someone moving very rapidly, relative to a stationary observer's clock. The most famous example is the *twin paradox*, in which a twin sister travels away from home in a spaceship at high speed, returning home to find herself younger than her twin brother. This prediction has now been tested experimentally many times—for example a comparison of the clocks on the Space Shuttle after it lands with those at home show that they do actually run slightly slower than those on Earth because of the high speed of the shuttle. The effect can be traced to the fact that events which appear simultaneous in one frame (e.g. separated flashes of light seen from one car, which is the frame) do not appear simultaneous when viewed from another frame (e.g. the same flashes seen from a different car travelling faster in a different direction), as we will discuss in more detail later. In Einstein's later general theory of relativity, published in around 1915, he was able to show that a strong gravitational field can also slow the passage of time. This was the context in which Einstein's theory appeared in 1905, which we will discuss in relation to the speed of light in Chapter 8.

There has been endless discussion about whether Einstein knew about the Michelson–Morley experiment when he wrote his famous 1905 paper on relativity, since he does not refer to that paper (or indeed any others). A good summary of the evidence is given in the book by A. Pais (1982). We have noted that Michelson–Morley does not show that the speed of light is independent of the speed of the light source, but it certainly calls into question the existence of an Aether, which could only be fixed to the Earth to be consistent with their result. This has a crucial bearing on the question as to how scientists actually work and make discoveries—the extent to which new theories depend on experimental evidence, and to what extent Einstein was influenced by knowledge of Michelson's experimental result in developing his theory of relativity. The question may be answered by Einstein's 1952 letter to R.S Shankland, with its echoes of the primary importance of Faraday's lines of force (**Figure 6.2**):

> I always think of Michelson as the artist in Science. His greatest joy seemed to come from the beauty of the experiment itself, and the elegance of the method employed. But he has also shown an extraordinary understanding for the baffling fundamental questions of physics. This is evident from the keen interest he has shown from the beginning for the dependence of light on motion. The influence of the crucial Michelson-Morley experiment upon my own efforts has been rather indirect. I learned of it through H.A. Lorentz's decisive investigation of the electrodynamics of moving bodies, with which I was acquainted before developing the special theory of relativity. Lorentz's basic assumption of an Aether at rest seemed to me not convincing in itself, and also for the reason that it was leading to an interpretation of the Michelson-Morley experiment which seemed to me artificial. What led me more or less directly to the special theory was the conviction that the electromotive force acting on a body was nothing else but an electric field ... There is, of course, no logical way leading to the establishment of a theory, but only groping constructive attempts controlled by careful considerations of factual knowledge".

Since Einstein had read Lorentz's 1895 paper, which discusses Michelson's experiments in detail, it is clear that Einstein was familiar with Michelson's work in 1905, but also that it was not the foundation of his theory. Rather, Einstein was driven by the need to find the simplest theory which would reconcile all the evidence, including Maxwell's prediction for the speed of light, Lorentz's transformations, and Bradley's result. His result had pleasing symmetries in the mathematics, and kept the form of the physical laws for both mechanics and electricity unchanged in different moving frames. As the title of his paper shows, Einstein was greatly influenced by the fact that, in Maxwell's equations describing Faraday's induction, one obtains the same result (but with quite a different interpretation) if a magnet is moved through a stationary coil, or vice versa; only the relative motion mattered.

The laws of electrodynamics prior to Einstein's 1905 paper (e.g. the Lorentz force law) use *the velocity of the electron*, and it was reasonable to ask "velocity with respect to what?" (a detector? the Aether?). This was important, because an electron at rest in one frame, and so not producing a magnetic field, would be seen as being in motion (and so producing a field) when viewed from another—his theory clarified all of this. But according to Abraham Pais, in 1905, Einstein was not aware of the "Lorentz transformations" already published by Lorentz in 1904 and, for a special case, by Voight in 1887—he derived them himself from first principles. Nor was he aware of Poincaré's work. (Einstein does say he was familiar with Lorentz's work in the Shankland letter, previously mentioned, written long after 1905). His biggest struggle was dealing with time dilation, which did not come to him until the very end, following discussion with his friend Michele Besso, the only acknowledgement or reference in his paper.

In his 1905 paper Einstein does mention "the failed attempts to detect the motion of the earth relative to the "light-medium" (Aether)," and he fully referenced all relevant work, including Michelson's, in a review article on relativity he wrote in 1907. Four months before his death in 1931, Michelson invited Einstein to Cal Tech for a dinner in Einstein's honor, where they met for the first time. In his speech, Einstein said

> *You, my honored Dr. Michelson, began when I was just a youngster. It was you who led physicists into new paths, and through your marvelous experimental work paved the way for the theory of relativity.*

Later in the evening, Einstein asked Michelson why he had put so much effort into his tedious measurements over so many years. "Because I think it is fun" said Michelson, in German.

It is therefore clear that both length contraction and time dilation had been presented in publications before Einstein's 1905 paper. And while most physicists were very reluctant to give up on the Aether drag hypothesis, they were unable to reconcile Bradley's aberration of starlight (no complete drag) with the Michelson and Morley result (complete drag). All, that is, except for Lorentz, whose 1895 paper, incorporating both the length contraction and time dilation ideas, was able to account for *both* the aberration and Michelson results. Yet Lorentz clung to the idea of an Aether, in which clocks told the "true time," while he was nevertheless a lifelong supporter of Einstein's theories. Poincaré also returned to support of the Aether in the years following Einstein's 1905 paper.

In the final 1950 edition of his authoritative, comprehensive, and mathematically detailed *History of the Theories of Aether and Electricity*, the eminent mathematician Sir Edmund Whittaker includes chapters on "The Age of Lorentz" and "The Relativity Theory of Poincare and Lorentz," in addition to chapters on Maxwell and Faraday, but none on Einstein, whose work he otherwise treats fully. For Whittaker, the importance of Einstein's paper lay in only two of its original contributions—his treatment of the relativistic Doppler effect, and of stellar aberration. While Whittaker's first edition of 1910, which does not cover relativity, is widely regarded as a masterpiece, he remains at heart a mathematician like Poincaré. Physicists have felt that he failed to recognize the physical significance of Einstein's paper, in which he derived the effects of time dilation and length contraction for the first time from first principles, leading to completely new equations for the energetics of colliding bodies, different from those of Newton.

Hendrik Antoon Lorentz went on to win the Nobel prize in 1902 for his work on the Zeeman effect (the effect of a magnetic field on atomic spectra), and for his introduction of the *Lorentz force* which acts on a charged particle moving in a field, as now taught in all undergraduate physics courses. He was awarded the Copley medal of the Royal Society, of which he was a Foreign Member. Unlike Einstein, Lorentz had led a quiet, highly successful scholarly life with a settled and happy family in Holland. After a PhD on Maxwell's equations in 1875, he was appointed Chair of Theoretical Physics at the University of Leiden where he stayed until 1912. With his wife Aletta he had two sons and a daughter Gertruda, who also became a physicist. He served as Director of the National Gallery of Fine Arts in the Netherlands, for which he designed their first postage stamp. After the First World War he led many administrative efforts to rebuild the country, including flood control with the Dutch dams and associated problems in hydrodynamics. The letters of European physicists from around the turn of the century make it clear that he was perhaps the most highly respected of them all, with the later exception of Einstein. Einstein and Lorentz spoke highly of each other throughout their lives. He died in 1928, with Einstein and Rutherford providing orations at the very grand funeral. You can watch a

black-and-white movie of the funeral from 1928 on YouTube at http://
www.youtube.com/watch?v=H2VtrJD0xJk.

Einstein had written of Lorentz, "For me personally, he meant more
than all the others I have met on my life's journey...I admire this man
as no other, I would say I love him.

8

Einstein

The Great Clarification

By 1904 things were in a complete mess. On April 27, 1900, Lord Kelvin gave an address at the Royal Institution, entitled "Nineteenth-century clouds over the dynamical theory of heat and light," citing the Michelson–Morley experiment and the black-body radiation problem as the two great unsolved problems in physics. The solution of the first led to Einstein's relativity; the solution of the second led to the birth of quantum mechanics. By this time the ageing Kelvin had become very opinionated, and had worked on the problem of the energy balance between light emitters and absorbers. He was well known not to be a good listener, unlike Rayleigh and Stokes. J.J. Thomson said of him, in this regard, that "he was a counter-example to the idea that a good emitter is a good absorber."

Many apparently crazy theories had been proposed to deal with the Michelson–Morley result—time slows down and things get shorter if you go really fast—at that time an ad-hoc hypothesis. An *emission* theory had been proposed, which always measured light speed relative to the source, which therefore discarded Maxwell's equations. Stoke's pitch-pine Aether with its velocity-dependent viscosity fitted well for the wrong reasons. In 1913, Willem de Sitter analyzed the light from a double star (one orbiting the other) to see if the speed of light was greater when one of the stars was coming toward the Earth, and slower when it moved away. It wasn't. But he did see the expected Doppler shift, which argued forcefully against the emission theory.

The complexity of the situation arose partly because, unlike the old action-at-a-distance theory, in which gravitational forces acted instantaneously across the universe, in the new field theory originated by Faraday, electromagnetic (and all other) forces took time to propagate at the speed of light—eight minutes and twenty seconds from Sun to Earth. During that propagation time, bodies will have moved, greatly

Lightspeed: The ghostly Aether and the race to measure the speed of light. John C. H. Spence.
© John C. H. Spence 2020. Published in 2020 by Oxford University Press.
DOI: 10.1093/oso/9780198841968.001.0001

complicating calculations. A sudden change in Saturn's orbit would not affect Earth's orbit until over an hour later.

In summary, at around the turn of the century, physicists were trying to reconcile the following results:

1. Maxwell's equations, which provided a constant velocity of light, suggested an absolute reference frame, supporting a stationary Aether through which the Earth moved.
2. The violation of the Galilean transformation for light. Unlike waves on a river, the speed of light waves did not seem to add to the speed of the Aether "current."
3. Michelson's experiment—no stationary aether, possibly Stoke's "complete drag."
4. The aberration of starlight—no "complete drag." No tilt of a telescope is needed if the Aether is fixed to Planet Earth.
5. Excellent agreement of several measurements of Aether drag with Fresnel's theory.
6. Newton's laws were independent of inertial frame under Galilean transformation, but Maxwell's were not.

Einstein's 1905 paper on "The electrodynamics of moving bodies" clarified and reconciled all these issues at a stroke, by incorporating both seemingly crazy ideas, abolishing the Aether entirely, and claiming that the speed of light was a constant (given by Maxwell's value), independent of the speed of its source. With no Aether, and a constant speed of light, the result of Michelson's experiment was immediately explained. But sorting out this mess was an achievement of genius—whereas the symmetries in Maxwell's equations and the results of the Michelson–Morley experiment supported his relativistic ideas, the results from the aberration of star-light and Fizeau's finding that the speed of light depended on the speed of a moving water medium, were at first harder to understand. The agreement with Fresnel's theory turned out to be a fortuitous agreement with Einstein's relativity theory, which results from an approximation linear in velocity in Einstein's correct theory. Einstein's paper contained a derivation of a new relativistic transverse Doppler effect, and a relativistic treatment of the aberration of starlight using the correct velocity addition law. The breakthrough came with his new understanding of the relativity of simultaneity and time dilation, which reconciled all these results, and led to a completely new understanding of the nature of time itself. Einstein realized that time intervals are measured by the coincidence of events, but these depend on the relative velocity of observers.

Albert Einstein was born in Ulm, Germany, in 1879 to Herman Einstein, an engineer, and Pauline Koch. In 1880 they moved to Munich, where his father and uncle established a company manufacturing direct-current (DC) electrical equipment. When the city requested bids for an AC electrical system, Herman was unable to respond, forcing him to sell the business and move to Italy. Albert moved later, in 1894, where he wrote a note (at the age of fifteen) entitled "The state of the Aether in a magnetic field." He was a difficult and rebellious student, strongly opposed to rote learning, but excelling in mathematics and physics. At the age of sixteen he failed the entrance exams to the Zurich Polytechnic, but was accepted at a second attempt a year later, and awarded a teaching diploma in 1900. He married in 1903 the only woman of the six students studying mathematics and physics for the diploma in his year, the Serbian Mileva Maric. Their daughter Lieserl, born in 1902, either died in infancy or was adopted. In that year he commenced work in Bern (now as a Swiss citizen) at the Swiss Patent Office. A son Hans Albert was born in 1904, and a second son Eduard in 1910, who suffered from schizophrenia. Einstein was awarded his PhD from the University of Zurich in 1905 for his work on Brownian motion. Mileva and Albert were divorced in 1919, and in the same year he married Elsa Lowenthal. He made his first trip to the USA in 1921, where he was struck by the "joyous, positive attitude to life...friendly, self-confident, optimistic and without envy." On a second trip to the USA in 1930 he visited Caltech, where his talk included the comment that science was inclined to do more harm than good. He also went to Hollywood, where he formed an immediate friendship with Charlie Chaplin. On his return to Belgium following a trip to the USA in 1933 he learnt that has house had been raided by the Nazis and his sailboat taken. He then renounced his German citizenship, and on seeing the growing harassment of Jews in Germany, travelled to the UK to ask Churchill for help to bring German Jewish scientists to Britain, which Churchill did. Einstein also helped to arrange for another thousand German scientists to move to Turkey. In 1933 Einstein migrated to the United States with Elsa. He remained at the Princeton Institute for Advanced Study, where he befriended Kurt Godel the famous mathematician, until his death in 1955.

At the age of twenty-six, in 1905 he published three of his greatest scientific papers, on relativity, on the theory of the photoelectric effect, and on Brownian motion. His paper showing the equivalence of mass and energy appeared the following year. He was appointed to a professorship in theoretical physics at the University of Zurich in 1908. Einstein's general theory of relativity appeared in 1916, and the confirmation in 1919 of its prediction that light rays would bend when passing a star due to its gravitational field brought him a life of celebrity and a host of honors. It is this theory which also predicts that the speed of light may change in a strong gravitational field. His Nobel Prize in 1921 was awarded for his work on the photoelectric effect, in which he introduced the photon, or quantum of light. In spite of also being a major contributor to the idea that sound and heat waves (phonons) also consist of quanta of energy, in later life he was never happy with quantum mechanics, as his famous ERP paper shows, which we discuss in Chapter 10.

Einstein was a socialist and strong supporter of international institutions who actively supported the advancement of African Americans, and an admirer of Gandhi. He supported many Zionist causes, including the establishment of the Hebrew University of Jerusalem, but turned down the offer to become President of Israel. He considered himself a "deeply religious non-believer" who did not believe in either a personal God or the afterlife. While he had written to the American President in 1939 to urge development of nuclear weapons (before Germany did), he otherwise spoke out strongly against war, and described himself as a pacifist, who believed that "without ethical culture, there is no salvation for humanity."

Einstein once said that if he had not been a physicist (who revolutionized the concept of time itself), he would have been a musician, and that he got most joy out of life from music. He played the violin, inspired by Mozart's music particularly. He played chamber music regularly in Germany, once playing with Max Planck and his son, and with the Julliard quartet in Princeton. The joke is told that another member of the quartet was once asked about Einstein's violin playing. "Einstein ?" he answered "He's OK. But he doesn't have much sense of time".

At this point we must take a brief detour from our main theme of the measurement of the speed of light, in order to understand Einstein's theory. As the title of his paper suggested, this was motivated by the problematic application of Maxwell's equations to light propagation in a moving medium, the problem which had preoccupied Michelson, Fizeau, and many other great nineteenth-century physicists. An understanding of Einstein's theory of relativity will lead us to a greater appreciation of the importance of the speed of light in calculating the enormous energy released by nuclear fission, with its well-known role in nuclear power stations and nuclear weapons.

There are two constants which repeatedly occur in the theory of relativity, so we may as well define them now since we will need them later. Relativity problems are generally concerned with observations made from two different frames of a fast moving object such as a meteor. These two frames could be someone on the ground, and someone in a car travelling at very high speed. We denote the speed difference between the frames as v (the speed of the car). Since about 1920, all physicists have used two symbols to represent this speed, firstly $\beta = v/c$ (the quantity directly measured by James Bradley in Chapter 4), which represents the fraction of the speed of light the car is going, and $\gamma = 1/\sqrt{(1 - \beta^2)}$, because this quantity occurs frequently in the theory, and it gets bigger when things go faster.

Einstein made two bold postulates in his 1905 paper. The first was that unlike all other kinds of waves, such as sound waves and ocean waves, the speed of light in a vacuum is constant for all observers, independent of the speed of its source or the detector. Thus there is no special frame of reference for absolute rest in which the true speed of light can be measured. The speed of the light from a car headlight does not change when the car speeds up.

His second postulate concerns the idea that if you are drifting steadily along in space (at constant speed) with nothing else around you, you should not be able to tell that you are drifting. Or even tell that you are moving. Or, in a spaceship, to get different results from simple physics experiments (like playing snooker on a table) from someone else whose spaceship is moving in a different direction at constant speed. This theory did not deal with acceleration—certainly if someone gives the table a push it will mess up your game (the snooker balls will run all over the table), and you can tell the difference from someone in another spaceship who was not pushed. Moving frames, like the interior of an elevator or a train moving at constant speed, are called *inertial frames* if they move at constant velocity with respect to each other, or, more precisely, if Newton's laws hold within them. In summary, Einstein claimed that the laws of physics should work the same in all inertial frames. The laws he had in mind were Newton's laws and Maxwell's equations. He required that these equations take the same form in any inertial frame. "Take the same form" means that if, for example, the equation for the force on an electron is proportional to the electromagnetic field in one frame, it should also show that it is proportional to the field when that equation is applied in another inertial frame (after allowing for their relative velocity), and not show that it is proportional to the square of the field, for example.

The idea that the laws of physics should be the same in all inertial frames had been suggested from the time of Galileo, so that transforming from one to another has been termed a "Galilean transformation." The basic idea is, again, that you cannot tell if you are moving, that is, moving with constant speed. "Cannot tell" means that no experiment in physics can tell you if you are moving with constant velocity. Here is Galileo himself discussing the issue. (We would now say *accelerating* where he says "fluctuating this way and that.")

> Shut yourself up with some friend below decks on some large ship, and have with you some flies, butterflies, and other flying animals. Have a large bowl of water with some fish in it; hang up a bottle that empties drop by drop into a wide vessel beneath it. With the ship standing still, observe carefully how the little animals fly with equal speeds to all sides of the cabin. The fish swim indifferently in all directions; the drop falls into the vessel beneath; and in throwing something to your friend, you need to throw it no more strongly in one direction than another, the distances being equal; jumping with your feet together, you pass equal spaces in every direction. When you have observed all these things carefully (though there is no doubt that when the ship is standing still everything must happen this way), have the ship proceed with any speed you like, so long as the motion is uniform and not fluctuating this way and that. You will discover not the least change in all the effects named, nor could you tell from any of them whether the ship was moving or standing still.

Galileo was correct—if we take Newton's laws and apply them to the emptying bottle and thrown object on the stationary ship, and then apply them on the same ship moving at constant speed in still water, they will make the same prediction for the motion of the droplets and the thrown object, as we would see on the moving ship. In Chapter 4 we saw how the speed of a person walking on a moving walk-way or belt at an airport relative to the ground is just equal to their speed relative to the belt, plus the speed of the belt relative to the ground. Provided the belt speed is constant, when extended to three dimensions (in vectorial notation), this mathematical relationship is the Galilean or inertial transformation and it transforms your speed relative to the belt to that relative to the ground. If you were somehow to play a game of snooker on the belt, you'd expect the balls to behave exactly as they would on a snooker table sitting on the ground, *provided* that the belt moved at a smooth constant speed, and did not accelerate, jolting the balls around. The collisions of the balls in snooker are described by Newton's three laws of mechanics, as are the motions of the planets.

This idea, that *all* the laws of physics (including both mechanics and electrodynamics) should apply (take the same form) in all inertial frames was one of the two fundamental concepts behind Einstein's theory of relativity. Partly because of their constant speed of light in any inertial frame, Einstein understood that Maxwell's equations did not obey this Galilean transformation, rather they obeyed a new "Lorentzian" transformation. This transformation coverts the coordinates on one frame to those in another inertial frame, but takes correct account of both length contraction (in the direction of motion) and time dilation, by introducing the factor γ into the transformation equations. By requiring this to also apply to Newton's laws, Einstein (aged twenty-six) came up with revised laws for Newtonian mechanics, which then held correctly at the very high speeds of nuclear particles, where Newton's original laws were found to fail. Lorentz had published his new transformation the year before Einstein's 1905 paper, which does not refer to it. Einstein derives it in an entirely different way.

We mentioned previously that two events (such as flashes of light) observed from one frame, such as a moving automobile, in which they appear simultaneous, may not appear so from another frame, such as a different car going at a different speed in a different direction. We now need to understand this in more detail, because this simple idea really lies at the heart of Einstein's theory, and if you can understand it—we will go slowly—I can assure you that there is nothing more complicated or difficult in the entire theory of special relativity. All the rest consists of developments and extensions of this idea, leading irrefutably to the craziness of time dilation (time slowing down as you go faster) and length contraction (things getting shorter the faster they go). So understanding the following is important—if you do, you can impress friends with your knowledge of Einstein's theory of relativity!

Figure 8.1 shows a fast-moving train car containing a lamp at its center, which radiates a flash of light in all directions at one instant. For

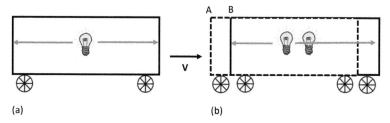

Figure 8.1   The relativity of simultaneity.

a passenger under the lamp in this car, the light from this lamp will be found to reach the ends of the car (where it magically rings a bell at each end) at the same instant—both bells will ring simultaneously in **Figure 8.1 (a)**. Remember, the light does not "pick up" the speed V of the train, but travels outward at the constant velocity given by Maxwell in Einstein's theory. However, for an observer on the ground listening from the side as it flashed past, the train in **Figure 8.1 (b)** would have moved forward a little from A to B during the time light was travelling to the front and back of the car. So the light would have less distance to travel to the back, and further to go in the forward direction. It follows that the person watching on the ground beside the train would hear the bell from the back of the car before the one at the front. In other words, an event which was simultaneous in one frame (on the train), was not, when viewed from another frame (the ground). In the dark, the person on the train has every right to believe that light reached both ends of the car simultaneously, *because he or she cannot know that he or she is moving!* (once we accept that there is no absolute "stationary" frame of reference provided by an Aether). This is the heart of the argument—the principle of relativity says that you cannot tell if you are moving (at a constant speed) by doing physics experiments.

The really important thing to understand is that these are not *appearances* due to a false impression created by movement—both observations are equally valid. For the person on the ground, the light really did hit the back of the train first. For the person on the train, it hit both ends at the same time. The problem lies, as Einstein realized, with our conception of the nature of time, which is not absolute, but depends locally on relative velocity. *Notice how a time interval of zero between events (the light arriving at the ends of the car) for the person on the train has changed into a finite time interval between these events for someone on the ground.* And there is no error introduced by ignoring the time taken by sound from the bells to reach the observers. We can get the same result if we take full account of this time, which is not important to the result—any kind of light detector could have been used, and the travel time of signals from these detectors can be allowed for in the analysis, but will not change the results.

In practice, this means that if you are flying from New York to Boston at high speed, and if below you the street lights of both cities are turned on simultaneously (in the ground reference frame) you will see the Boston lights turned on first. So once we start dealing with very high relative speeds (as in astronomy or nuclear physics), time intervals themselves become dependent on speed and location, and this led Einstein to re-formulate his theory in terms of a four-dimensional mixture of space and time.

These considerations raise apparent conundrums when we think of cause and effect. It would seem obvious that an effect cannot precede its cause. If the occurrence of one event before another, or the other way around, depends on the velocity of the observer, we might ask how this relativity of simultaneity affects the idea that one event can cause another. To do so the causal event must happen first. But does relativity say that we could zoom past in some direction with a particular speed which makes an effect appear to precede its cause? It turns out that this depends on whether there is time for light to travel from one event to the other. If there is not enough time for light to travel between them, they cannot be causally connected and can appear in any order. If there is, then they will appear in the correct sequence in all frames of reference. If you throw one rock at another, causing it to move, your rock must have taken longer than light would to get there. Since it not possible to throw a rock faster than the speed of light, you could not cause the second rock to move by throwing the first rock faster than light.

This result has many profound consequences, and also leads to many paradoxes, described and resolved in undergraduate physics textbooks, which lie outside the scope of this book. The extension of these ideas, when quantified, leads directly to the extraordinary idea of time dilation, that moving clocks run slow by an amount γ, as derived in Appendix 4. That prediction has been confirmed experimentally by comparing an atomic clock on an intercontinental airliner after many flights with an identical clock which remained stationary on the ground. The aircraft's clock was found to be slightly slow—it had ticked fewer times than the clock on the ground. The time difference was about a tenth of a microsecond. The same results were found in an experiment involving the International Space Station. As we approach the speed of light, these effects become larger. If the "away" clock travels at half the speed of light ($\beta = 0.5$, $\gamma = 1.154$), then, after an hour's travel, on its return the ground clock will show that 1.154 hours have passed.

We find these predictions of length contraction and time dilation surprising only because we don't normally zoom about at sufficiently high speeds to notice the effects. Many of the most fundamental discoveries in science are due to scientists trying to imagine what happens at extreme scales—very big things (the universe and the big bang), very small things (quantum mechanics), very fast things (relativity), very long time periods (Darwin's theory of evolution), very short time periods (atoms bonding together), very large energies (supernova star explosion), and so on, all unfamiliar to our human experience on a much more limited scale of time, distance, speed, and energy. Knowing one scale of time, about 4.5 billion years for the age of the Earth, has been absolutely crucial to the acceptance of Darwin's theory of evolution, since this age was believed to be much less during Darwin's lifetime, and not long enough for new species to evolve, as we will briefly discuss in Chapter 10.

These predictions of Einstein's regarding time dilation have important practical implications—our GPS systems must allow for the effects of special relativity, which introduces a correction of seven microseconds per day. Critical GPS daily corrections are needed to prevent accidents between self-driving cars and can only be predicted by assuming that the speed of light, bringing news of distant events, is constant, independent of the speed of the source or detector, and that there is no Aether or absolute frame of rest.

In addition to time dilation, Einstein's paper predicted the length contraction (in the direction of motion) mentioned in Chapter 7. So with a modified clock and a change in length, it is not surprising that the Galilean rule for the addition of velocity vectors had to be changed to satisfy relativity. In Chapter 4 we obtained the Galilean rule by thinking about a person walking along an airport beltway. If the beltway is travelling at velocity $\mathbf{v}$ with respect to the ground, and the person is walking with velocity $\mathbf{u}'$ with respect to the beltway, then the person's velocity $\mathbf{u}$ with respect to the ground will simply be sum of $\mathbf{v}$ and $\mathbf{u}'$, or

$$\mathbf{u} = \mathbf{v} + \mathbf{u}'.$$

When the effects of length contraction and time dilation are included, Einstein found that this Galilean rule was changed to

$$u = (v + u')/G,$$

where $G = 1 + (vu'/c^2)$. Unless the velocity of the beltway v or the person's velocity on the beltway u' approach the speed of light c, G is about unity, and the two equations give almost the same result, as used in Newtonian physics. But for the aberration of star-light problem (as in Chapter 4, where v becomes the velocity of the Earth around the Sun, about 70,000 mph), a correction to the predicted angle measured by Bradley is needed, and this was provided by Einstein. Using the new *velocity addition law*, Einstein was finally also able to explain both Fizeau's light speed measurements through moving water, and give the correct relativistic Doppler effect.

Einstein's new addition law makes an interesting prediction. Even if we run at the speed of light on the beltway (so $u' = c$), then our speed relative to the ground becomes $|u| = c$, and never exceeds the speed of light. This demonstrates the profundity of Einstein's theory—it means that no *object* can exceed the velocity of light. This result has nothing to do with light, just the requirement for Lorentz invariance, which in turn is traceable either to the desire for mathematical elegance in our theories which best fit the facts, or, physically, to the relative simultaneity problem. In spite of its counterintuitive prediction, it happens to be true! It is also possible to see, using $E = \gamma mc^2$, that it would require an infinite amount of energy E to accelerate a mass m to the speed of light, since γ is infinitely large when $v = c$ and $\beta = 1$.

Einstein's 1905 paper also dealt with changes needed to Newton's laws as a result of relativity. At that time, Newton's laws for mechanics did indeed take the same form after a transformation from one frame or coordinate system moving at constant velocity to another. However, with the development of nuclear physics in the early twentieth century, it was becoming clear that Newton's laws were failing, or incomplete, for these processes, when the particle velocities became very large.

In addition, Lorentz had shown in 1904 that there was a form of linear transformation under which Maxwell's equations would transform correctly, so that these equations retained their form in a different frame. These transformations generate the length contraction and time dilation effects, which are obtained by extending our argument about simultaneity being relative. Einstein had derived them in his paper from first principles (see Appendix 4). In fact, Lorentz had shown that simply the requirement that Maxwell's equations retain their form in different inertial frames was sufficient to define the required transformation, which has thus become known as a Lorentzian transformation.

The quantities which must be transformed are a mixture of electric and magnetic fields in four-dimensional space, so that in relativistic electrodynamics, the amounts of each depend on the velocity of the frame, and electric and magnetic fields lose their individual identities as they become mixed together.

In four dimensions, Maxwell's four equations reduce to just two, and this insight was an important motivation for the development of relativity. It is remarkable that Maxwell's equations, derived from a mechanical model, are relativistically correct. We need to express them in the four dimensions of space and time to see that clearly. In three dimensions, physicists would say they had a "hidden symmetry."

In summary, there were then three possibilities for the future of physics at the end of the nineteenth century:

1. Find the Aether. Retain Galilean transformations for mechanics (Newton's laws), but for electrodynamics (Maxwell's equations), find evidence for an absolute frame of reference in the universe, in which light has the speed Maxwell obtained in terms of other fundamental electrical constants, avoiding the need for a relativity principle. Locate the absolute rest frame experimentally, as Michelson and Morley had attempted to do.

2. Fix Maxwell's equations. Show that Galilean transformations apply to both mechanics and electrodynamics. The fact that Maxwell's equations change form when transformed in this way just means they are wrong and need to be repaired.

3. Fix Newton's equations. Show that the same transformation law (e.g. Lorentz transformation) can be found for both electromagnetism and mechanics, and the fact that Newton's laws change form when transformed in this way just means that they are wrong, and need to be fixed up, especially at high speeds.

In view of all the experimental evidence summarized in Chapter 7, and for reasons of mathematical elegance and simplicity, Einstein decided that the last option was the best, and so derived new equations for mechanics. These only agree with Newton's laws if things are going slowly compared with the speed of light. This was a very bold and imaginative thing to do for a twenty-six year old, to say the least, given the authority of Newton. For example, Einstein showed that Newton's most famous equation, $\mathbf{F} = m\mathbf{a}$, giving the force $\mathbf{F}$ on a mass m which would give it an acceleration $\mathbf{a}$, had to be changed to $\mathbf{F} = \gamma\, m\mathbf{a}$, to satisfy Lorentz invariance. With these new laws, moving clocks run slow, and fast rulers get shorter in the direction of their motion. Perhaps most importantly, in the next year, Einstein published the equation

$$E = m\,c^2,$$

giving the amount of energy E which can be released in a nuclear explosion or nuclear power reactor by annihilating an amount of mass m in a nuclear reaction. This occurs when nuclei break up and atoms are split. The speed of light is a very large number, so a very small mass loss generates a huge amount of energy in, for example, the nuclear reactions which occur in our Sun, which is a continuously exploding atom bomb. The first experimental confirmation of this equation occurred in 1938, immediately before the outbreak of the Second World War, when Hahn and Strassmann measured the release of energy when the uranium nucleus was split by bombarding it with neutrons. You can read the full story of the consequences of this discovery for the development of the atom bomb in the superb account by Richard Rhodes (1986).

Einstein obtained this equation by looking for a new definition of momentum (mass m times velocity $\mathbf{v}$). According to Newton's laws, momentum is conserved, and has the same value before and after particles collide. Einstein needed a new kind of relativistic momentum which takes account of relative velocities, and would be conserved in particle collision, but would also preserve the conservation law when transformed from one frame to another by the Lorentz transformation.

The new relativistic definition of momentum required is $\gamma m\mathbf{v}$. By applying this to an inelastic collision of particles (like lumps of clay thrown together and sticking) or other methods, we obtain the most famous equation of physics, $E = mc^2$. You may get a headache trying to apply this equation to the quantum particles of light (photons), since they have zero mass, but not zero energy. In terms of the "rest mass" m_0, this equation reads $E = \gamma\, m_0\, c^2$, and only by making the speed of the particle $\mathbf{v}$ tend to the speed of light c (so γ gets very big) can we prevent the energy going to zero as the mass goes to zero. Maxwell's equations give the correct energy of light as its momentum multiplied by the speed of light.

The reluctance to accept Einstein's theory of relativity is easy to understand. As Michelson (who eventually completely accepted relativity and its predictions) complained in his book around 1927,

> *The existence of an Aether appears to be inconsistent with relativity . . . but without a medium, how can the propagation of light be explained?*

It was said that it was easier to understand the mathematics of special relativity than the physics of it, and very difficult to accept replacements for Newton's laws, due to his immense authority. As one professor in physics has commented "in physics, mathematics can easily be a substitute for thought," a view which Thomas Young wrote strongly in support of. Sabine Hossenfelder has written a delightful and amusing book about the exaggerated importance given to the "beauty" criterion for new equations in the subatomic high-energy particle physics community. This was never the case for Einstein, especially as a young man, for whom physical intuition always came first. In his later years he did turn increasingly toward more formal mathematical manipulations in his unsuccessful pursuit of his unified field theory.

Let's look back at the whole story in this book now, and try to provide the view from Mount Olympus of this entire intellectual journey of ideas. Action at a distance, the idea that gravity and light acted instantaneously across the universe, held sway until the time of Roemer and Bradley, who provided the first strong experimental evidence for a finite speed for light. These measurements were vital in helping to provide a scale for the universe and solar system. Competing explanations for refraction threw into sharp relief the disagreement between those who thought light was a wave, and those favoring a particle model; the "corpusculists" versus the supporters of "undulations" in an elastic, invisible Aether which somehow could not support longitudinal waves.

Next, Thomas Young showed that light, split into two beams can be recombined to produce interference fringes, in exact accordance with a wave theory of light. Faraday, the great experimentalist, was the catalyst for many major theoretical insights. He saw in his iron filings tensioned lines of force, along which waves might travel, giving birth to field theory and a finite velocity for light. If his iron filings, shown in **Figure 6.2**, were indeed shaped like elliptical grains of rice, balancing on a side, I can easily imagine that they would jiggle at the slightest touch to the apparatus, suggesting tension and hence a finite speed for a pulse running along them. His discovery of the Faraday rotation of the polarization of light led Maxwell to the concept of his displacement field. Later, this helped to explain how radio waves are propagated, as FitzGerald was the first to understand. Magnetism, electrostatics, and optics were at first considered entirely unrelated phenomena, to be unified by Maxwell the magician, with his mechanical model of the Aether, later discarded, and his demonstration that light was an electromagnetic wave, for which he provided a constant speed (in terms of electrical constants) which did not depend on the speed of the source of the light. This added support for the existence of a frame of absolute rest in the universe (the Aether) which supported the propagation of light waves.

Fresnel's Aether drag theory (not to be confused with frame-dragging in general relativity) supported experiment for a century, while the brilliant terrestrial measurements of Fizeau, Foucault, and Michelson both improved on the accuracy of lightspeed measurements and addressed the problem of light propagation in a moving medium. This culminated in Michelson's null result, which gave the same speed for light in all directions on a moving Earth.

Rayleigh (and Gibbs and others earlier) clarified the distinction between group and phase velocity, indicating that all measurements except Bradley's had been group velocity measurements, the speed at which information is transmitted, not that of the crests of the waves. Poincaré and Lorentz anticipated many of Einstein's 1905 results but retained the idea of an Aether. Einstein finally wrapped it all up and clarified everything in 1905 in a theory which also, as a result, could extend Newton's laws to the very high energies and speeds of nuclear physics and predict the energy release from atom bombs. His theory connects space and time through the speed of light.

9

Radio and Telecommunications
Spacecraft

Some time ago I attended a scientific conference filled with space-craft communication engineers. During a presentation on the *Cassini* spacecraft mission to Saturn, I was astonished to learn that that the radio transmitter used on the spacecraft to communicate with Earth had a power of only fifty watts, less powerful than a small incandescent reading lamp. How can this weak transmitter possibly provide communication across the hundreds of millions of miles between Earth and Saturn, and how long does the signal take? In this chapter we will try to understand how radio waves can propagate over such vast distances in a complete vacuum, travelling as they do at the speed of light.

The *Cassini* space probe, which was a collaboration between USA and Europe, was launched in 1997, and is shown in **Figure 9.1**. It had a 4 m high-gain parabolic antenna at the top and a radioisotope thermo-electric powered radio transmitter, with travelling-wave vacuum tube amplifier, and ran faultlessly for twenty years. *Cassini* had originally been planned for only four years, but it eventually spent thirteen years orbiting and photographing Saturn—it really is worth a look on the NASA/ESA/ASI website to see the stunning results. Finally, it passed through Saturn's rocky rings and was crashed intentionally into the surface of the planet in 2017, allowing scientists to confirm Maxwell's theory of the stability and formation of those rings. Maxwell had provided a theory to explain the stability of the rings, explaining why the rocks in the rings don't either crash into Saturn, or fly off into space.

Cassini carried the small *Huygens* module, which landed on the surface of Saturn's moon Titan in 2005, sending back priceless data and pictures for ninety minutes after landing. Titan, a rocky moon discovered by

Lightspeed: The ghostly Aether and the race to measure the speed of light. John C. H. Spence.
© John C. H. Spence 2020. Published in 2020 by Oxford University Press.
DOI: 10.1093/oso/9780198841968.001.0001

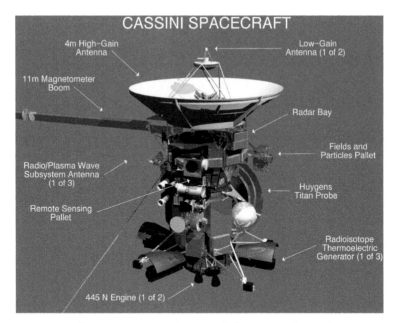

Figure 9.1 The *Cassini* unmanned spacecraft. (Courtesy of NASA.)

Huygens in 1665 and 40% of the size of the Earth, is extraordinary in that it has an atmospheric pressure about one and a half times greater than that on Earth, composed mainly of nitrogen, which allowed the Huygen's probe to be parachuted onto the surface.

The distance from Saturn to Earth is 746 million miles at closest approach, across which they had to communicate. With the speed of light about 671 million miles per hour, radio waves take about 746,000,000/671,000,000 or one hour and seven minutes to get to *Cassini*, so if NASA send a control signal ("*Watch out for that asteroid!*") nothing happens for over an hour! And that time is longer when Saturn is farthest from Earth. But then the picture of the threatening asteroid previously sent back from *Cassini* to Earth would also have arrived an hour late, so when NASA responded, it would have been far too late anyway! In fact, as Maxwell had predicted, the rocks in Saturn's rings are actually very far apart, so there was little chance that *Cassini* would hit one.

In previous chapters we have seen how the electric or magnetic fields which so fascinated Faraday are simply defined as a force on a charge or current. The key to generating radiation (radio waves) from these charges is to make them accelerate, for example by oscillating them back and forth along a wire causing the fields to radiate with the frequency of the oscillation, as shown in **Figure 9.2**. But can we really believe that some electronic charges, sloshing about along a short wire antenna near Saturn from a fifty watt transmitter can impose a force from a field, on charges millions of miles away, in a similar radio antenna on Earth? How is this possible?

The first point is that these antennae are resonantly tuned to each other, in just the way a guitar string vibrates when you hold a tuning fork near it, when they are tuned to the same pitch (frequency). The other strings don't vibrate, so any extraneous noise is excluded. This helps a lot. Secondly, the "X-band" frequency used around 10 GHz corresponds to a wavelength of only three centimeters, allowing fairly small dish antennae (like those used for satellite TV) to direct the radiation into narrow cones pointing at each other. The angular width of this beam is limited by the diffraction effects discovered by Fresnel and Rayleigh. For *Cassini*'s 4 meter diameter dish at Saturn transmitting across the 746 million miles to Earth, the width is about 0.5 degrees. This angle was first given in a famous formula by Rayleigh, which also tells us the finest detail which can be seen under a microscope, telescope, or camera. In this case it says that the angular width of the beam

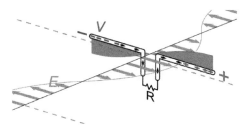

Figure 9.2 Dipole antenna receiving a radio signal. The green arrows are the electric field arriving, which provide the force on charges (electrons) in the wire, causing them to move and so creating the alternating electric current shown by the black arrows. This charges the two sides of the antenna rods alternately positive and negative, causing a current to flow through the resistor R, which is amplified by the receiver. By making the length of the rods equal to half the wavelength, a standing wave is set up, as for a resonant guitar string and tuning fork. The same antenna can be used to transmit radiation by acting in reverse. On *Cassini*, a similar radiator is mounted at the center of the high-gain antenna, shown in **Figure 9.1**, and the 3 cm radio waves are focused by the "dish" toward Earth. (Courtesy of Chetvorno, Wikimedia Commons.)

is proportional to the wavelength of the radiation divided by the diameter of the dish, so a bigger dish makes a finer beam. Here, it means that the beam spreads to a width of about 5.5 million miles when it gets to Earth an hour later. The fifty watts of power transmitted by *Cassini* are spread over this 5.5 million mile diameter disk. So the dish on earth can capture only a tiny fraction of this. For this purpose, NASA supports three seventy meter diameter dishes at its Deep Space Network near Los Angeles (Goldstone, near Barstow USA), Canberra (Australia), and Madrid (Spain). If this signal can be detected, in addition to providing communications, the Doppler shift on the radio signal from *Cassini* can be used to monitor its velocity relative to the Earth.

A famous formula due to Dr H.T. Friis was published in 1946, which gives the power received by an antenna due to a transmitter some distance L away, in terms of the size of both antennae, the distance between them, and the wavelength. The article by Shaw (2013) gives a full explanation of this equation. With today's intense competition over spectrum space and real-estate for mobile phone antennae, this equation has taken on the greatest importance, as customers complain about "no service" indications when they move out of range of a cell-phone tower. An interesting question the equation can answer is whether the towers would cover a greater range if they operated at higher or lower frequency. These mobile phone towers transmit in the gigahertz range on similar frequencies to the deep space communication with *Cassini*.

We can use the Friis equation to estimate the signal strength on Earth from *Cassini* near Saturn. The equation for the power received P_r in terms of the power transmitted P_t at wavelength $\lambda = 3$ cm is

$$P_r = P_t \left[A_r A_t / (L^2 \lambda^2) \right],$$

where A_t and A_r are the areas of the transmitting and receiving dishes. We can use 70 m for the Goldstone receiving dish on Earth, 4 m for *Cassini*'s transmitting dish, and 746 million miles for the distance L between Saturn and Earth. We then find the power received on Earth from *Cassini*'s 50 watt transmitter to be about 2×10^{-15} watts, or a thousandth of a millionth of a millionth of a watt! This is all the power available to move those electrons up and down the antenna rod in **Figure 9.2**. In 2014, the *New Horizons* spacecraft managed to communicate at a thousand bits per second from Pluto, where messages travelling at the speed of light take 4.5 hours to reach Earth, using only twelve watts of power at 10 GHz. What is important is how these signals compare with other unwanted signals in the same frequency band, coming from the same direction—the background or signal-to-noise ratio. We can make a rough estimate of this background signal.

The air we breathe consists mainly of nitrogen and oxygen molecules separated by many times their diameter by vacuum, in which they zoom about at several hundred miles per hour, colliding with each other and with solid objects. Because these collisions are so numerous, they share their energy about equally. In addition to his research on Saturn's rings and electromagnetism, Maxwell worked out the average speed of these molecules, while the average energy of a molecule in a gas was later shown by the German physicist Ludwig Boltzmann to be proportional to the temperature T, with a proportionality constant k named after him.

Similar arguments can be made for the electrons in the metal rods of a dipole radio receiver antenna, which also share this amount of energy kT. This has the value of about twenty-five millielectron volts of energy on a mild day. This energy, due to the thermal jiggling of the electrons in wires, produces the main background signal which will compete with the incoming transmissions from *Cassini*. It produces the hiss from a radio which is not tuned to a station. Lightning and the cosmic microwave background are additional sources of background. As might be expected, the signals from *Cassini* will be easier to read if they are sent slowly, as for ordinary speech. The ability to distinguish the signal from the noise therefore also depends on the rate or frequency Δf at which the binary bits or Morse code are sent. The product $kT\Delta f$ is known as the noise power, and it sets a fundamental limit on the performance of the super-sensitive radio receivers at NASA deep space communications facilities around the world.

To reduce the background, we need to cool the input electronics of the receiver, to make kT smaller by reducing the temperature T. This can be done using liquid helium, which boils at about four degrees above the absolute zero of temperature on the Kelvin scale (4 K). This scale is the same as the centigrade scale C, except that it sets the lowest temperature possible to zero, which would correspond to -273 °C, or -459 °F. Liquid helium can be produced just below its boiling point using a sophisticated refrigerator. We may be willing to accept the binary bits coming in as slowly as a hundred thousand per second, corresponding to a bandwidth of $\Delta f = 100$ kHz. Then if we cool our receiver to 4 K, we will have a background noise "floor" of about $kT\Delta f = 5 \times 10^{-18}$ watts for the background hiss. By comparison, the bandwidth of a modern wi-fi system is about 20 MHz. The important point is that this background power is less than *Cassini*'s signal, so it should be possible to detect *Cassini* even at the enormous range of Saturn's orbit. Of course this is a very simplified analysis, and there may be many other sources of background, such as a cook in the Goldstone facility restaurant starting up a microwave oven, which leaks radiation—microwave ovens work on similar frequencies. A more sophisticated calculation, which included other sources of noise, brings the actual signal-to-noise ratio down to about unity. Using various digital tricks, modern GPS receivers, such as the one in your mobile phone, can detect signals which are even weaker than the noise, if the digital "bits" are sent slowly enough.

Next, we need to understand radiation—does all of the field produced by a moving charge propagate outward forever if it is not focused by a dish? Certainly it will get weaker, as the same amount of energy is spread over a sphere of ever increasing radius. The static electric field discovered by Coulomb and the magnetic field discovered by Biot–Savart decay with distance from their source rapidly, the energy in their product declining inversely as the *fourth* power of distance from the source. As long as the charges responsible don't accelerate, they don't radiate, and they only produce this static field.

We can use Maxwell's equations to solve for the fields generated around the simplest oscillating dipole antenna, as Hertz was the first to do, with charges oscillating and now accelerating along a wire as in **Figure 9.2**. We then find that the energy in the field in a particular direction decays inversely as the *square* of the distance from the antenna, not the inverse fourth power. This much stronger field is the radiation field. When the total power radiated is added up over the surface of a sphere around the antenna, the result is a constant, independent of the size of the sphere, since, when added up, all the radiated power must cross the sphere, however large it is. But the amount of power intercepted by a small patch on the sphere of constant size, or by a small antenna of fixed size, decreases inversely as the square of the distance, even if a focusing dish is used.

Radio waves were discovered by Heinrich Rudolph Hertz in 1887 in a brilliant piece of experimental physics which could serve as a model for all scientists. It could be said that he also founded the telecommunications industry. We have seen that Maxwell's equations were understood to explain the propagation of light, but the idea that they also described electromagnetic waves at lower frequencies emerged only slowly after Maxwell's death. This radiation was at first called "electromagnetic light." Hertz's aim was to see if these electromagnetic waves could be generated, and, if that was possible, to measure their speed. If this was equal to the speed of light it would confirm their similarity to the light waves described by Maxwell's theory.

Heinrich Hertz was born in 1857 and as a student, proved equally gifted in the humanities and sciences. He delighted in the design and construction of scientific instruments such as galvanometers. He became a pupil of Helmholtz in Berlin, where he soon stood out by solving the problem Helmholtz had given him, of showing that a moving electrical charge in a conductor has inertial mass. Hertz had a complicated relationship with his mentor, the great Helmholtz. While grateful for the support of his boss, who always had good ideas for him to explore, Hertz was keen to develop his own ideas and seek independence—the sign of a good student. In 1879, through the Berlin Academy of Sciences, Helmholtz offered a prize to any scientist who could "establish experimentally any relationship between electromagnetic forces and the dielectric polarization of insulators." Helmholtz suggested this research project to Hertz for his PhD. Hertz concluded that any effect would be undetectably small, and devoted his thesis instead to the induction of charged rotating spheres. He thereby demonstrated his abilities as a first-rate applied mathematician, in addition to his experimental skills. He completed his PhD in about a year, with the highest honors, remaining as Helmholtz's assistant in Berlin for three years, where he published thirteen papers on various subjects, including work on cathode rays. Cathode rays are the flow of an electron current in a vacuum. Hertz was working at a time long before the electron was discovered, by J.J. Thomson in 1897.

Hertz moved to Kiel in 1883, then to Karlsruhe as Professor of Physics in 1885, and finally to Bonn in 1889 as the successor to the famous physicist Clausius, one of the founders of the subject of thermodynamics. Hertz was plagued by ill health throughout his life, and died from blood poisoning on January 1, 1894 at the young age of thirty-seven.

Helmholtz died eight months later, aged seventy-three, having made major contributions to many fields. He was one of the rare European physicists who had studied Maxwell's work. However the action-at-a-distance school from Neumann and Weber, which taught that light and electrical forces travelled instantaneously, was the dominant intellectual tradition in Germany at that time. Helmholtz had devised a compromise theory, which he taught to Hertz, in which forces travelled instantaneously in free space, but not within a medium. In Helmholtz's theory, electrical disturbances propagated more slowly by the mechanism of polarization. This means that electricity travels in a substance by changing the shape of each atom slightly as the effect moves along a chain of atoms bonded together.

In his first appointment away from Helmholtz, alone and miserable in Kiel, Hertz published a paper in 1884 in which he re-derived Maxwell's equations from first principles. He used a novel approach, avoiding mechanical analogies and Maxwell's model for the displacement current. The paper is remarkable for deriving the modern form of Maxwell's equations purely in terms of fields, not potentials, as Heaviside was also to do. These equations were referred to as "Maxwell's equations in the Hertz-Heaviside form." Hertz always gave credit to Heaviside for these equations, however in Europe his own influence became dominant.

Hertz's subsequent 1890 paper on "The fundamental equations of electromagnetism" further developed Maxwell's equations in abstract form and became the foundational paper for the study of electromagnetism in Europe. Arnold Sommerfeld, a founding theoretician for quantum mechanics in the 1920s and earlier, said of this paper "the shades fell from my eyes, and I understood electromagnetism for the first time."

As a result of Helmholtz's influence, Hertz became preoccupied with finding out if electromagnetic disturbances travelled at the same speed in matter as in free space or air. In doing so, he was to initiate a long tradition of measurements of the speed of light using radio waves. At Karlsruhe, Hertz inherited a great deal of useful equipment, including Ruhmkorff induction coils, which had been invented in 1836. They were the first form of high voltage AC transformer and generated a large electrical spark. Similar to the ignition points that interrupt the current to the spark coil in old automobiles, Ruhmkorff coils had a mechanical interrupter to rapidly stop and start the voltage from a battery, creating a crude type of alternating current which fed into a small primary coil with few turns, and a large secondary coil with many turns.

The process by which rapidly changing current in the primary coil transfers energy to the secondary coil is electromagnetic induction, which was discovered independently by Michael Faraday (1791–1867) in England and Joseph Henry (1797–1878) in the USA. Inductance is a property of all coils. One Henry is the amount of inductance necessary to induce one volt when the current in a coil changes at a rate of one ampere per second. One Farad is the capacitance in which one coulomb of charge causes a potential difference of one volt.

In 1853 Kelvin showed that if a coil (inductor) is coupled in parallel with a capacitor (an LC circuit), a process of resonance would occur in which energy could be transferred back and forth between them. He was able to show that the frequency (f) of this resonance was defined by a formula which is now known to all electrical engineers as

$$f = 1/(2\pi \sqrt{(LC)}),$$

where C is the capacitance in Farads and L is the associated inductance of the coil in Henrys. The values of L and C can be calculated from the dimensions and number of turns of the coil, and from the size of the capacitor, and accordingly, the resonant frequency of an LC circuit can be calculated.

To elicit resonance of any sort requires some form of excitation. For example, a bell is hit with a hammer and a guitar string is plucked. The striking action generates a wide range of frequencies, of which only those which are close to the resonant frequency will be picked up. How can resonance be excited in an LC circuit, making it oscillate? What is required is a rapidly varying current.

A spark was an ideal source because it generates a very wide range of frequencies. As already mentioned, sparks could be generated from Ruhmkorff coils or by the discharge of capacitors made by metal foil separated by insulating glass (Leyden jars). In 1847 Helmholtz suggested that the discharge of a capacitor would generate an oscillatory current.

In a series of experiments between 1887 and 1889 Hertz analyzed the waves produced by spark-excited resonant circuits, and in so doing, he demonstrated the first intentional generation of electromagnetic waves (radio waves). By measuring their wavelength and combining it with their frequency calculated from Kelvin's formula, he was able to calculate their speed using the formula $v = f\lambda$. He found them to travel at the speed of light, and like light, to satisfy Maxwell's equations.

Hertz's experimental progress toward generating radio waves through a series of complicated experiments is a fine example of a scientist groping in the dark (literally and figuratively), guided by theory, and just trying things out. Since the very existence of radio waves was not known, we cannot be surprised at his mis-steps and the many blind alleys he followed. Remember that the distinction between Faraday's near-field induction effects (the way a transformer sends energy between two unconnected coils with a gap between them) and true radiation as we saw for *Cassini* was not clear, even in the theory, at this time. Hertz's approach was always admirably systematic, making slow and incremental progress while he built on things that began to work. For example, he had to find the best kind of sparks to use, the best spark-gap material, and the best orientation for his detector, while recording it all in his notebook as the months went by. Along the way he discovered the photoelectric effect, in which ultraviolet light shining on the spark-gap electrodes made the sparks bigger. Judging from his letters, papers, and books, his research program is probably the finest example of systematic experimental research in this book. It demonstrates exactly the tenacity and patience required of a superb experimental physicist, all the time guided by a rich hinterland of theoretical understanding.

Hertz pointed out that his discovery could not have occurred by theoretical predictions alone. There simply was no theory to suggest that a spark-gap could be used as a detector of radio waves, as he finally realized it could. This realization, more than anything else, accounted for his success—in his day there were no devices that worked as a radio receiver. The train of experiments leading to his crucial one is so complicated, with so many blind alleys, that I will have to greatly simplify it. In 1887 he made the crucial observation that a spark in one location in his laboratory would sometimes produce a spark across an identical gap and circuit some distance away. The second gap was not connected to a battery, and there were no wires between them (*Wire-less!*). The challenge for Hertz was to understand why. He also had to distinguish this effect from Faraday's induction, which only worked over short distances, and did not involve the propagating electromagnetic waves which Hertz discovered. In textbooks we often "tidy up" a sequence of very complicated historical events, re-writing history to make it easier to remember as a dramatic story. No doubt there is some evolutionary "fitness" or "survival value" operating in this tendency, which helps to

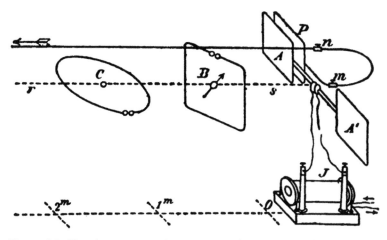

Figure 9.3 Hertz's experimental arrangement for measuring the strength of the electromagnetic waves in a wire using a spark gap detector. (From Hertz (1893).)

preserve memories. This is particularly true in the case of Hertz's work, which certainly did not have the aim of "discovering radio waves."

Hertz was not the first to notice these jumping sparks from "open circuits," but at that time, with FitzGerald, he was one of the very few people able to interpret the phenomenon in terms of electromagnetic theory. No doubt people had noticed random sparks occurring in a room during a lightning storm. The high-voltage Wimshurst machine also became popular in Victorian England around this time for party demonstrations, in which sparks ran from hand to hand between guests standing in a circle. In his first experiments, Hertz's coupled resonant circuit oscillated at about thirty-three megahertz (nine meter wavelength). The unit of frequency has since been named after him, with one Hertz being one cycle per second. A big advance occurred when he realized he could greatly increase the size of the spark in the detection loop by making the detector resonate to the same frequency as the transmitter circuit. He did this effectively by using the same values of inductance and capacitance in both circuits. He also got stronger sparks in the receiver by working at shorter wavelengths, an effect which FitzGerald had predicted, as we will see. These shorter wavelengths at higher frequencies provided him with wavelengths which had the considerable advantage of fitting into his room for his standing wave measurements.

Figure 9.3 shows Hertz's diagram of the apparatus for his early experiments. The induction coil at J supplies a high voltage to the "transmitter" spark-gap above it. We could think of the plates A and A′ and the upper wire loop as forming an antenna. Much of Hertz's work consisted in trying to figure out the radiation pattern of this antenna, and how his measurement of it depended on the orientation of his detector loops B and C. These are resonators ("receivers") each with its own spark-gap in different orientations and positions. The spark-gap distance could be adjusted using a micrometer, and he measured the strength of the electric field in space from the length of the spark it generated, for example near B. Modern attempts to reproduce his work have shown how difficult it must have been, closing down the gap at the receiver in a pitch-black room until a tiny weak spark first appeared, then recording the result in his laboratory notebook. Between A and r in his laboratory, the waves, whose length scale is given in meters below, ran along the wire and were picked up by the detector loop. On November 7, 1887 he was first able to observe that the tiny spark at B, say, in the receiver would come and go as he moved that wire loop away from the "transmitter" at A in the direction of the arrow.

On November 9, 1887 his wife Elizabeth wrote to his parents that he had

> . . . *again succeeded in the most beautiful experiments, which . . . make him very happy, and me as well, when he tells me about it with such a radiant face.*

Much more experimentation followed. By March 1888, he was finally able to map out the maxima and minima of radio-frequency standing waves excited in air throughout his laboratory, as shown in **Figure 9.4**. Today you can repeat his experiment in your kitchen. Microwave ovens heat food using radio-frequency standing waves. Appendix 5 explains the nature of these waves, and shows how you can make them in pizza dough in a microwave oven and determine the wavelength of the microwaves by measuring the distance between crests. The frequency of the microwaves is written on the back of the oven. From the known frequency and measured wavelength of the waves you can calculate their speed, just as Hertz did to establish that they were electromagnetic waves travelling, like light, at the speed of light, but at a lower frequency. As later pointed out by Poincaré, Hertz made an error in estimating the frequency of his radio waves, which he calculated to be 177 MHz. This resulted in a value for the speed of light which was too

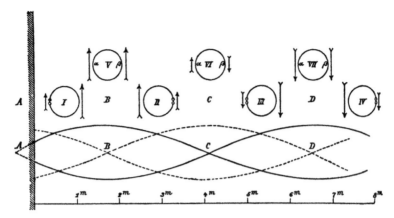

Figure 9.4 Hertz's diagram showing how he measured the nodes and antinodes of radio standing waves in his laboratory. These standing waves behave just like a vibrating string on a piano, where the ends are fixed and immobile, forcing them to become nodes and fixing the wavelength and frequency. (From Hertz (1893).)

large. Poincaré corrected it, to give excellent agreement with Maxwell's value. You can find full details needed to repeat this experiment with simple modern equipment in the article by Faccio and Clerici (2006).

In retrospect, Hertz's three great ideas which made the discovery of radio possible were the use of a barely visible spark in the dark as a radio-wave detector, his realization that resonant circuits would increase their strength, and his idea of using standing waves in a room to measure their wavelength. His original aim had been to test Helmholtz's hypothesis on the nature of electromagnetic disturbances in matter, by comparing their speed of transmission in matter and in air, and so to test the instantaneous transmission idea. In so doing he was led to the discovery of the existence of propagating electromagnetic waves in air at frequencies lower than that of light—radio and microwaves.

Hertz went on to undertake many experiments at a shorter wavelength of 66 cm, thereby discovering microwaves. He demonstrated that the waves were transverse, and showed that they obeyed the laws of geometric optics when their wavelength was much smaller than that of diffracting objects. Returning to Helmholtz's problem, he studied the transmission of radio waves through media, including large masses of pitch and wood. A later paper from 1889 entitled "The forces of electrical oscillations, treated according to Maxwell's theory" became hugely influential throughout Europe and Britain. This was the first mathematical treatment of radiation from a dipole antenna, exactly like a modern television antenna. His wife Elizabeth drew the maps showing the field lines around the dipole, according to Hertz's solutions of Maxwell's equations, showing much more fine detail than that sketched schematically in **Figure 9.2**.

Hertz was unlucky to die young, before his discovery bore fruit in the development of commercial radio just ten years later. However, his discovery was widely recognized throughout Europe and especially in Britain, which he visited in 1890 to receive the Rumford Medal of the Royal Society. As Johanna Hertz records, "he wrote to his family, with the exasperation of someone whose well-meaning hosts left him with little time for himself to explore London and do as he pleased . . . ":

> I had a very good time in England . . . it was not sheer pleasure but in many respects a great strain . . . This time I was seasick both ways; I am evidently getting old. Professor Ayrton . . . had invited me to stay with him . . . he had taken rooms for the four days for himself, his wife and me in the Langham Hotel, in order to do the honors of London for me. That was not at all what I had wished, but having once accepted I could not very well change anything; moreover they both acquitted themselves of their task in the most amiable manner. On Saturday afternoon he took me for a walk through London, and to the Royal Institution, where Faraday had worked. Professor Dewar . . . showed me everything. He mentioned that they would like to have me there for a lecture as well, he thought my English was more than adequate. In the evening we sat around the hotel fireplace and talked. Sunday morning we first attended services at Westminster Abbey, and then looked in on a small art collection (I really would have preferred to walk through the city, but in that regard I was somewhat at a disadvantage, in that I was never left alone, never without a guide). At one o'clock there was a lunch at the hotel, attended by 15 physicists, among whom you would recognize the names of Crookes and Lockyer, dear Papa. This lunch turned into a reception . . . with about 70 people, most of whom I knew, for example Swan, who introduced the incandescent lamp before, or about the same time as Edison; and Hughes, the inventor of the microphone and the automatic telegraph. In the evening, Prof Ayrton, Professor Lodge and I were guests of Prof Lockyer at the Reform Club, a magnificent building in Pall Mall; it was well worth while to get a glimpse of these clubs. On Monday I first gave some thought to my speech for the evening. (In the afternoon he visited several laboratories in South Kensington). At 3 o'clock we all went to a meeting at the Royal Society. Here I was of course especially fascinated to meet the older of the foreign physicists, Sir W. Thomson, Sir G. Stokes, Lord Rayleigh and others who had all come to the meeting. Acceptance of the medal required merely a bow on my part, and no speech, which I found much to my taste. I did notice that my coming in person made a very favorable impression. The dinner began at 7 pm. The speeches did not start till after the dinner, and by our lights were dreadfully long, all prepared in

advance and managed by a toastmaster, a completely new custom to me. The Lord Chancellor toasted the Royal Society, Sir William Thomson the medalists, Dr Hopkinson responded for the medalists in general, I for myself in particular. Of course I made it short. I had read what I wanted to say over for Prof Ayrton beforehand, and received nothing by flattering comments afterwards. In any case, I think I did better than if I had spoken German. The Italian ambassador, who responded for the guests in Italian, apologized most handsomely for not speaking English, but no one understood him, and having to apologize in this way made a rather foolish impression. After dinner . . . I was taken (or dragged) by a chief engineer to an underground station for electric lighting, where no less than 1200 horsepower was converted in a small cellar room. I was actually rather tired, but in retrospect was rather glad to have seen these things . . . Tuesday morning, until 1 o'clock I spent in the National Gallery, which does contain some very magnificent paintings, many more than I remembered. Then I was Professor Lodge's guest at the Liberal Club, the noblest ornament of which is Gladstone, in a completely new building on the Thames Embankment, and really grandly furnished. Since Prof Ayrton had a meeting, I was at liberty for the first time, and took a two and a half hour walk through the City, from London Bridge to the Tower. I would not care to live in London, but to walk through it is interesting enough. At 6 pm we had dinner with Professor Lodge and FitzGerald of Dublin, who had not previously attended the meeting and whom I now became acquainted with as well. Then we all went to the Lyceum theatre to see Ravenswood, based on Scott's The Bride of Lammermoor, a piece of inanity in my opinion and badly acted, although the actors and the audience thought quite otherwise . . . On Wednesday I got up early and travelled to Cambridge, where I met four or five more Professors in related fields, saw the Physics laboratories, Colleges, Newton's relics . . . arrived here (in Germany) yesterday at 1 o'clock and found everything in the best of order, was greeted by my students here with just as lively demonstrations as at the Royal Society on Monday, so that I really could not ask for anything better.

Hertz's work, with its measured value for the speed of propagation for electromagnetic waves, ended support for the action-at-a-distance instantaneous propagation theories of Neumann, Weber, and Helmholtz. Hertz went on to do much more influential research before his early death in 1894. He measured the resonance curve of an electronic oscillator circuit for the first time, analyzed the skin effect in detail, and built the first coaxial cable and "slotted line" waveguide, which found a practical application more than fifty years later for radar.

In Bonn he hired Philipp Lenard as his assistant for more work on cathode rays passing through thin metal plates. Lenard was later to win the Nobel Prize (first awarded in 1901) for this work in 1905, which formed the basis for Roentgen's discovery of X-rays in 1895. Lenard used cathode rays hitting a metal plate in an evacuated tube to produce X-rays. An important part of Hertz's legacy was his calculation of the pattern of radiation for a dipole antenna, and the dependence of its radiating power on frequency. This proved of fundamental importance for Max Planck's papers in around 1900, which established the idea that energy is quantized, coming in small indivisible lumps. Planck got the result by treating a hot radiating cavity (a "black body") as if it were lined with millions of these tiny dipole radiators, matching the emission spectrum to experimental data. Under equilibrium conditions, they only agreed if the energy of the dipoles came in discreet amounts, like water waves which can only be exactly an integral number of centimeters high. It has also been argued that Hertz's paper on "The fundamental equations of electromagnetism," which assumed that the Aether moved with material as a temporary assumption, provided essential background for Einstein's theory of relativity. Einstein, Planck, and Lorentz all had a high regard for Hertz's work. He died at the age of 36 in 1894.

Hertz's work established the tradition of measuring the speed of light using radio waves. In the 1930s, the Russian physicists Mandelshtam and Papaleksi made very accurate measurements of the speed of light by transmitting radio waves of a few hundred meters' wavelength to a reflector and timing its return to the transmitter. The returned signal shows a phase lag with respect to the transmitted signal, which can easily be measured. To improve accuracy, they detected the signal at the reflection site and re-transmitted it back at a different phase-locked multiple of the original frequency, then studied the phase difference using Lissajous patterns on an oscilloscope. Their result for the speed of light, finally obtained in the midst of the Second World War, was the most accurate to date. In retrospect, their experiences can be seen as the beginnings of radio astronomy and radar.

There are many stories about the unwitting detection of radio wave effects prior to Hertz's discovery in 1887. In England, Lodge, Heaviside, and FitzGerald were closely involved in developing theoretical and experimental methods, but they were clearly scooped by Hertz because they could not invent a suitable detector. In around August Lodge had started to think about experimental methods for generating "electromagnetic light." This included a scheme for frequency doubling with a series of capacitors to make visible light, and another, anticipating Hertz's discovery, based on the oscillatory discharge of a capacitor. All of these schemes prior to Hertz foundered on the problem of building a detector. Some of them undoubtedly worked to make radio waves, but the waves could not be detected. Bolometers (heat detectors), resonators, and standing waves were all considered. As we have seen, Hertz eventually first detected them by using the very faint spark they created in a nearby spark-gap, connected to a resonant circuit.

After first publishing a paper in 1879 showing that electromagnetic wave generation was impossible using electrical forces, in which he ignored displacement currents, FitzGerald eventually reversed this conclusion. Soon after, he was able to develop a mathematical treatment for the standing wave field around an oscillating radio-frequency current in a wire, as suggested in **Figure 9.2**. At the time, the relationship between the excitation of the atoms of a medium such as the Aether and light propagation was not understood, since this required a theory of electron motion, and the electron was not to be discovered until 1897. Light was still seen by most as a "bodily undulation of the Aether" quite different from any electromagnetic waves. In around 1882, however, FitzGerald, stimulated by Rayleigh's work on sound, became convinced of the reality of electromagnetic wave propagation at frequencies lower than that of light. He focused on calculating the energy carried away by radiation, and on developing the concept of retarded potentials. This idea, that fields and potentials travel at finite speed, so that the Maxwell equations must allow for a delay, had been suggested independently in the work of L.V. Lorenz (not H.A. Lorentz) in 1867, of which Fitzgerald seems to be unaware. FitzGerald's later publication of 1883 used this concept, and thereby assumes that radio waves do travel at finite speed, the speed of light. Remarkably, he gives the correct expression for radiated power as a function of frequency for the first time. This power depends on the fourth power of the frequency. The result later became of the greatest importance in the research leading up to the birth of quantum mechanics around 1900 in Planck's analysis of the radiation from a "black body," mentioned earlier. FitzGerald's result allowed him to find the frequencies and wavelengths where significant energy would be radiated. In an abstract for the September 1883 *53rd British Association for the Advancement of Science Meeting*, he wrote

On a method of producing electromagnetic disturbances of comparatively long wavelength. This is by utilizing the alternating currents produced when an accumulator (capacitor) is discharged through a small resistance. It would be possible to produce waves of as little as 10 meters wavelength, or less.

In this way, FitzGerald was opening up for the first time the entire electromagnetic spectrum for study, but it was the experimental approach of Hertz that made it a reality. The papers of Lodge, FitzGerald, and Heaviside in the late 1880s, based on Maxwell's equations, thus laid many of the foundations for the theory of radio-wave generation. It seems very unlikely that the young Heinrich Hertz was aware of much of this work, published in English, and, in the case of FitzGerald, in obscure Irish conference reports. These "Maxwellians" became the heroes of British science when Hertz's discovery was announced by FitzGerald at the British Association meeting in Bath in 1888. Until Marconi's work ten years later, it was assumed that radio waves would be no different than light waves, hence limited by clouds and useful only for line-of-sight communication, and therefore not much use.

Perhaps the most intriguing pre-discoverers of radio were Wilhelm Feddersen and David Hughes. Feddersen did generate radiation at about one megahertz in 1859, using the spark discharge from a Leyden jar capacitor, which was associated with some inductance, thereby generating damped electrical oscillations. He then tried to measure the speed of propagation of the radiation using rotating mirrors, as Wheatstone had done. However Feddersen did more than repeat Wheatstone's experiments, since he was measuring propagating waves, rather than an electric pulse running along a wire, and because he was perhaps the first to use photographic recording of spark images.

Another fascinating tale concerns the crucial difference between the short-range induction effect and the longer range propagating radiation terms we discussed in connection with the *Cassini* probe and the radio antenna in **Figure 9.2**. As we have seen, Maxwell established that light was an electromagnetic wave, but his writings contain no mention of radiation at lower frequencies, such as radio waves, which he could easily have predicted. The Welshman David Edward Hughes (1830–1900), now almost entirely forgotten, was a remarkably inventive scientist who could reasonably lay claim to have discovered radio-wave propagation in September 1879, just two months before Maxwell died and well before the discoveries of Hertz in 1887 and Marconi.

David Hughes was a gifted musician who had toured the music halls of London and performed for the Royal Family as a child with his brother and sisters as "child prodigies." Their show was so successful they were able to take it to the USA, where they performed at the White House. The family became comfortably wealthy and purchased a farm in Virginia in the 1840s. Here the boy began to experiment with chemistry and electricity in a laboratory built for him by his father. He first saw a telegraph machine as a teenager, and before long he obtained a position as Professor of Philosophy and Music at a college in Kentucky. Like Wheatstone, Hughes became obsessed with the development of a printing telegraph which would avoid the need to learn Morse code. He invented a machine into which letters could be typed directly, then encoded by electro-mechanical switching, transmitted, and reproduced at the other end. His system used a rotating cylinder rather like a musical box, and a similar rotating cylinder at the receiving end. Synchronization of the sending and receiving cylinders was provided by a vibrating reed. The machine was driven by a falling weight and did not require power to be extracted from the telegraph pulses themselves for synchronization and printing, nor did it need large batteries. This remarkable machine was capable of simultaneously transmitting and receiving in what is now known as *duplex mode*. It worked well, and by 1855 it could operate at forty-four words per minute and was patented and used in the United States.

An improved version, developed with the American Telegraph Company (ATC) was then prepared for use on the new Atlantic telegraph. In 1858 Hughes became involved in the extremely competitive UK telegraph market. After many misadventures and conflicts with the management and staff of ATC, his system eventually succeeded. It was first adopted in France, but in time became the standard system used throughout Europe for all international lines operated by the International Telegraphic Union, making him a wealthy man. It remained in use throughout the world until about 1940.

In 1877 Hughes moved to London. He was the first to coin the term microphone. With A.G. Bell's invention of the telephone, Hughes turned his attention to its improvement, and eventually he invented the carbon granule microphone, which was a big advance on Bell's magnetic diaphragm. It was based on the principle that sound vibrations on a diaphragm compress carbon granules between electrodes, increasing their conductivity. Hughes decided not to patent his invention, in order to facilitate its use and development by others, which rapidly followed. The result was a bitter dispute with Edison, who claimed priority. The dispute was finally resolved by Kelvin, who found that they had worked independently.

Unlike Bell's electromagnetic diaphragm microphone, carbon microphones had the enormous advantage that they produced large voltage signals, which was of very great importance because Bell's electromagnetic diaphragm microphones produced very weak signals. Receivers were based on exactly the same principles as Bell's microphones and had very low sensitivity, so the sound in telephones was very weak. The large voltage signals produced by carbon microphones meant that the sound was much louder and could overcome losses in long distance wires. The receivers remained unchanged for many decades, but carbon microphones were rapidly adopted. They were used in almost all telephones and communications equipment up to the 1980s, when low cost miniature electronic amplification became practicable, and they were gradually replaced by dynamic or crystal microphones which had greater fidelity.

Carbon microphones were important for other reasons. Because they produced such large voltage signals, they could be used as amplifiers for weak signals. This was done by acoustically coupling a standard electromagnetic receiver unit to a carbon microphone. In this way a weak signal could be regenerated, and "carbon amplifiers" were widely used in long distance telephone systems until well into the twentieth century.

It could even be argued that in 1879 carbon microphones led Hughes to discover radio transmission, long before Hertz, as a result of an accidental electrical spark that occurred when his new microphone lay nearby. By September 1879, just before Maxwell's death, Hughes was able to leave his apartment and walk 500 yards away carrying his "microphone", acting as a receiver, and still hear the clicks from electrical sparks back in his room, which were making radio waves. At this distance he must have been detecting propagating waves from wires acting as an antenna connected to his spark generator. Had he used a resonant coil and capacitor circuit on both spark generator and receiver, as did Marconi years later, the clicks would have been much louder.

The first glimpse of the mechanism by which Hughes' carbon microphone acted as a detector came some ten years later. In 1890, the French physicist Edouard Branly showed that a tube of conductive particles showed greatly increased conductivity in the presence of minute electrical charges, and in 1894, Oliver Lodge showed that Branly's "coherer detector" could be used to detect radio waves, far better than Hertz's spark-gap. The reason why the particles attract one another and so conduct electricity under an AC radio-wave field is not understand well even today. It may involve micro-welding between particles or quantum mechanical tunneling.

Hughes was greatly excited by his discovery and wrote extensive notes on it, which were subsequently published in the fascinating book by J.J. Fahie (1899) on the history of radio up to that time, which is still in print. Hughes repeated this demonstration for several senior members of the Royal Society, including Stokes. Unfortunately, Stokes interpreted the results as due to Faraday's well-known short-range induction effect, and declared that it was therefore not novel. Lodge or FitzGerald may have reacted differently! Bitterly disappointed, Hughes was to live to see the discovery and use of radio in the coming years before his death in 1900. Fortunately, the historical record has been maintained because William Crookes successfully persuaded Hughes to include an account of this research in Fahie's book and in technical journals after the discoveries of Hertz and Marconi had been announced.

Hughes made many other discoveries, including the experimental demonstration of what we now call the *skin effect*, whereby higher frequencies travel mainly at the surface of a conductor, the theory of which had already been worked out by Heaviside. Hughes invented the *twisted pair* cable wiring system to reduce interference on telephone lines, which is still used today in nearly all computer cables and networks. He invented many other electrical devices, including an induction balance.

In 1882 Hughes married, and in 1886 he became a Fellow of the Royal Society. He was awarded the Royal Medal and the Albert Medal, and became vice-president of the Royal Institution. He devoted much of the final years of his life to helping younger scientists at the London Polytechnic. He was described as mild mannered, generous, genial, and sympathetic, always full of interesting anecdotes and light-hearted and excellent company. You can read the full story of his life in the book *Before We Went Wireless* (2010). He left part of his estate to a London hospital fund which is still operating, and part for a medal to be awarded by the Royal Society. Recipients of the Hughes Medal have included Steven Hawking, Max Born, and Niels Bohr.

10

Faster-than-Light Schemes

Quantum Reality

We end our odyssey with a survey of developments following Einstein's work at the beginning of the last century, and the realisation that the speed of light is constant. These developments have included a new *definition* of the speed of light, several questionable schemes for transmitting information at speeds greater than the speed of light, and some paradoxes in quantum mechanics which appear to do so. And we should think again about some of the questions we asked at the start of this book—how do light or radio waves travel millions of miles through a complete vacuum, so that a vibrating atom in the Sun causes electron motion in our eye eight minutes later, and an impression in our brain of sunlight? And what exactly *is* an electric field in outer space—what stuff is it made of? And if we don't know, why does our mathematics describe and predict its behavior so well?

We've seen how rationalizing the definition of the units of magnetism and electrostatics in the nineteenth century led to Maxwell's value for the speed of light. This process of simplifying, improving, and reducing the number of independent units continued into the twentieth century, with important implications for the speed of light. We saw how Arago started out his career doing surveying in Spain to assist with the definition of the meter in 1793 as a fraction of the size of the Earth (one ten-millionth of the distance from the equator to the North Pole). From these measurements a length standard was defined by a special platinum ruler kept in Paris. The unit of mass, the kilogram, was originally defined as the mass of a liter of cold water, but later better defined by the weight of a particular chunk of platinum–iridium alloy metal, kept in a safe in Paris. Other nations sent their copies of this standard regularly to check that they had not changed. If the one in

Lightspeed: The ghostly Aether and the race to measure the speed of light. John C. H. Spence.
© John C. H. Spence 2020. Published in 2020 by Oxford University Press.
DOI: 10.1093/oso/9780198841968.001.0001

Paris lost a few atoms every year, it did not matter, since it was the standard!

The importance of international agreement on standards is far greater than is generally realized, not only for physics and chemistry, but also for trade, where fraud is common. The British and US gallons are different! Antigua and the USA continue to use the obsolete British imperial system, long since abandoned by the UK and the rest of the world.

In particular, the semiconductor industry may be the most demanding in their requirements on standards. Semiconductor devices are "grown" by evaporating atoms onto a silicon plate in a vacuum chamber, layer by layer, from a gas containing these atoms. This is done at many fabrication plants ("fabs") around the world, each plant costing as much as twenty billion dollars. For the latest devices, some of these layers are only a few atoms thick, so that measurement of the density of the gas (mass per unit volume) becomes critically important. Recently, it has been found that the standard of mass is only just sufficiently accurate to make reproducible devices. These must have exactly the same number of atomic layers in every device, whether fabricated in Ireland or Arizona. All the "fabs," running under highly automated control, must be synchronized by satellite communication. The flow rate of the gas is also synchronized and controlled by atomic clocks. Clearly, the world needed, long ago, a better standard of mass than the weight of a piece of platinum (an "artifact" or physical object) kept in Paris. And certainly a better standard of time than any pendulum or even quartz clock could provide.

It seems remarkable that Maxwell himself understood this need for permanence and reproducibility in standards as long ago as 1870. His own work on relating the standard units in electricity and magnetism had led him, via his current balance, to a measurement of the speed of light. Here is what he said in Liverpool in 1870 at the British Association for the Advancement of Science meeting that year on the question of standards, anticipating the modern trend away from the use of artifacts as standards:

Yet, after all, the dimensions of our Earth and its time of rotation, though relative to our present means of comparison, are very permanent, they are not so by physical necessity. The Earth might contract by cooling, or it might be enlarged by a layer of meteorites falling on it, or its rate of revolution might slowly slacken, and yet it would continue to be as much a planet as before. But a molecule, say, of hydrogen, if either its mass or its time of vibration were to be altered in the least, would no longer be a molecule of hydrogen. If, then, we wish to obtain standards of length, time and mass which shall be absolutely permanent, we must seek them not in the dimensions, or the motion, or the mass of our planet, but in the wavelength, the period of vibration, and the absolute mass of these imperishable and unalterable and perfectly similar molecules.

Another scientist working in Ireland at around this time was the eccentric George Johnstone Stoney (1826–1911). Based on Faraday's experiments on electrolysis, Stoney became convinced that there was a fundamental unit of electric charge, and correctly estimated its value. The electron was not discovered experimentally until 1897, by J.J. Thomson. Stoney is famous for giving the name "electron" to this unit of charge. History does not relate whether Stoney had read what Maxwell had to say (above), but in 1874 he presented a proposal for a new set of units based on a very small set of fundamental entities in nature, rather than man-made artifacts or human anatomy, such as the length of the Pharaoh's foot. His choice was his unit of electrical charge e, Newton's gravitational constant G, and the speed of light c. He then showed how these could be combined to give new units of length, mass, and time. For a quick example, the units of distance divided by those of time give us a unit of speed, perhaps in miles per hour or kilometers per hour.

Continuing this search for units which will not change for all eternity and in all places throughout the universe, Max Planck later proposed certain combinations of Newton's constant, the speed of light, and the constant h he discovered in around 1900, which is the fundamental quantum of angular momentum. It is remarkable also, that three of these constants can be combined to form a dimensionless number known as the *fine structure constant*, $\alpha = e^2/(2 \varepsilon_0 h c)$, which has the approximate value of $1/137$ in all systems of units. It occurs frequently in quantum mechanics. Here ε_0 is the dielectric constant in the SI system of units. The constant α also measures the strength of the electromagnetic interaction between charged particles, and so measures the strength of the binding between atoms. For a while, this constant was believed by some physicists, including Arthur Eddington, to have exactly this value. Then $1/\alpha$ would be equal to the integer 137, a source of great mystery and speculative "numerological explanations" as to why God might have constructed our universe in this unique manner, constrained by a magical integer!

More accurate measurements have shown that $1/\alpha$ is not an integer. However it does contain the speed of light, so the question arises as the whether the value of α has changed since the dawn of time, and hence whether the speed of light really has been the same since the big bang, or does this "constant" itself vary in time? To answer this question, we would need measurements of α soon after the big bang, to compare with today's value. And we'd need to assume that the charge on the electron and Planck's constant had not changed. This also leads to the more general question as to whether the laws of physics have changed over the age of the universe. Paul Dirac, in 1937, suggested that perhaps the constant in Newton's gravitational law has changed since the big bang.

In the 1960s, George Gamow and others hit on the idea that, since the light from atoms in distant stars comes from the distant past, it could be compared with light from the same atoms in a laboratory on Earth. It was known that a certain feature in the spectrum of the light, the *splitting of doublets*, had a separation which depended on the fine structure constant. A comparison of the separations would provide a way to compare the current value of α with the value a few billion years ago, using light from a distant star. The age of the universe since the big bang is 13.8 billion years. Since Gamow's paper, many physicists have worked to improve and develop this spectroscopic method, including a more sensitive approach which uses the absorption of light travelling to Earth through dust clouds. To cut a long story short, at first several measurements were published which showed a very small change in α over billions of years. John Webb's positive results in Sydney in 1999 attracted much attention at the time, as described in his article in *Physics World* (2003). However, the majority of recent results now show that α has not changed within error. Increasingly accurate measurements continue to be made. A search for variations in the magnitude of α with position in the universe is also an active field of current research. For a delightful survey of the relationship between all the fundamental constants in physics and their significance, I can recommend John Barrow's book *The Constants of Nature* (2002). This also summarizes the various *varying speed of light* (VSL) theories which have been proposed in cosmology.

It now seems fitting to review the modern definitions of some units, one of which has meant that it is no longer useful to measure the speed of light. Since Einstein, all scientists have agreed that the speed of light is a universal constant, perhaps the most important constant of all. Taking this value as our starting point, we can define other quantities in terms of this constant. For example, if I know that a car will certainly travel at exactly 100 mph forever (please excuse imperial units!), and I had a super-duper atomic clock of perfect accuracy, I could define the mile as the distance this constant-speed car will travel in one hundredth of an hour. I am *assuming* that the speed of the car is a known constant, and having defined the mile in this way, there would be no point in trying to measure the speed of the car. That is what was decided by the international standards committee in 1983. They decided that "the metre is the distance travelled by light in vacuum during a time interval of 1/299792458 of a second." Therefore the speed of light will forever after be exactly 299,792,458 meters per second. That also gets rid of one artifact—the platinum meter ruler in Paris.

For a time standard, one second was historically defined as 1/86400 of a day, clearly not a very good unit in view of the gradual slowing down of the Earth's rotational speed due to the tidal effects of the Moon. The current definition of a second is 9,192,631,770 times the period of certain microwave radiation from cesium-133 atoms at approximately the frequencies used by mobile phones, which is very close to the traditional length of the second. This standard has been used since the 1960s and is generally accurate to within one second in a few hundred million years. A set of these atomic clocks around the world "votes" on the current time, and the average is used for *International Atomic Time*. From this consensus, a *Coordinated Universal Time* is derived, equal to this, but with *leap seconds* added as needed to allow for the Earth slowing down. This time standard is broadcast on shortwave radio on 10 MHz and used by GPS systems for synchronized global timekeeping and its myriad uses.

Recently, at the US National Institute of Standards laboratory in Boulder Colorado, the accuracy of atomic clocks has been improved by better shielding from their environment, improved temperature control, and by using the faster oscillations from ytterbium atoms instead of cesium atoms. In this way a clock has been made which achieves the astonishing accuracy of not gaining or losing more than one second during the age of the universe (about fourteen billion years). Through Einstein's theory relating time dilation to gravitational field, this clock gives us a way to map out gravitational fields. It is so sensitive to changes in gravity that it can detect a change in its height of one centimeter (gravity gets weaker as you move away from the center of the Earth).

Finally, the last artifact, the kilogram mass, has also been eliminated in much the same way as the length standard, by defining the kilogram in terms of another universal constant, Planck's constant. This constant of angular momentum, with units of kilogram per second per square meter, contains the units of mass (and the other established constants of time and length), so if we assume that it is constant, we can use its relationship with time, distance, and mass to define the kilogram. A decision to do so was made by the General Conference on Weights and Measures in November 2018. At the same time, the units of current, temperature, and the mole (the amount of a substance) were re-defined in terms of other fundamental known constants, rather than using unique preserved artifacts. Designing a scientific instrument to "weigh" masses according to this new mass standard is another story! This can be done either by balancing a mass against a magnetic field, measured in terms of Planck's constant (a Kibble balance), or by counting the number of atoms in a perfect sphere of silicon.

So now all the standards of the international metric system have finally been re-defined in terms of fixed physical constants. In summary, the metric system of units, the quantity they represent, and the corresponding physical quantity they are defined by are as follows: kilogram (mass, from Planck's constant); meter (length, from speed of light); second (time, from cerium microwave radiation frequency); ampere (current, from charge on the electron), Kelvin (temperature, from Boltzmann's constant); mole (amount of substance, from Avogadro's constant); candela (luminous intensity, from luminous efficacy of light at a specific frequency and power per unit solid angle).

Although we've covered a lot of territory in this book, perhaps we are really nowhere nearer to understanding what *is* an electric field. In deep outer space, how does light propagate? Is nothing needed to support its propagation, since Einstein abolished the Aether? And has Faraday's idea of an electric field at a point in space—the force which a small charge at that point would experience—just replaced one kind of Aether (an elastic medium) with lots of properties, by another (a field) with fewer properties? To bring our story up to date with developments in physics since Einstein in 1905, we need to consider the *quantum field theory* that was developed soon after by Dirac, Jordan, Heisenberg, Pauli, Feynman, Schwinger, Dyson, and others. This theory predicts that there *is* a *background energy* which can never be removed, and is present everywhere, even in complete vacuum.

This new quantum field theory combined quantum mechanics with special relativity, and imposed the discreteness of the quantum on the energy in these fields. The quantized fields which result were shown to be capable of creating particles, consistent with Einstein's equation $E = mc^2$. Most remarkably, the theory predicts that the vacuum state is actually teeming with short-lived "virtual" particles such as photons and radio waves, as first predicted by Max Planck in 1912. In other words, you cannot take all the energy out of a vacuum, even at the lowest possible temperature, and that the last half-quantum cannot be removed. Such a notion is consistent with the Heisenberg uncertainty principle and allows that "virtual" photons can exist briefly without violating the energy conservation principle. Their average energy at zero temperature would give a gigantic energy contribution to an empty universe before the big bang. Fluctuations in this "zero-point" energy were once thought to be responsible for bringing our universe into existence. According to one estimate, the energy contained in every cubic centimeter of vacuum is about 0.6 electron volts. Einstein further developed this theory, and in 1920 referred to it as a new kind of Aether, which did not consist of "ponderable matter." This postulated zero-point energy of the electromagnetic radiation field leads to measurable effects, such as the spontaneous emission of light and the remarkable Casimir effect. The Casimir effect is a force due to the zero-point energy which exists between two metal plates that are brought very close together, facing each other, in the dark, in a vacuum and at the lowest possible temperature. It is due to an imbalance in these zero-point waves, which always exist in vacuum. A voltage is not applied between the plates.

The Casimir force has been measured experimentally. We can understand it by analogy with a similar force which actually occurs when two large ships pass each other, going in opposite directions. The water waves are akin to the zero-point radio waves of the vacuum state, and these span a very wide range of frequencies and wavelengths in the space outside the two ships, extending to the horizon. Between the ships, however, there can be no waves whose wavelength is larger than the distance between the ships, which will satisfy the wave equation and these boundary conditions. For each wavelength there is a corresponding frequency. In this water-wave analogy, the energy carried by these waves depends on their height and frequency. From the difference between the longest wavelength (and corresponding cutoff frequency) allowed for waves between the ships, and outside the ships, it follows that the total amount of energy stored by the wavefields is different in the two regions, because there are fewer long wavelength waves carrying energy between the ships than outside them. This difference in wave energy leads to an attractive force on the ships, sucking them dangerously together as they pass. Force is a measure of the change in energy with distance between the ships, just as the energy in a spring increases when it is compressed. This is a real effect, so be very careful next time you steer one aircraft carrier past another! The radio waves and infra-red waves between two metal plates in vacuum have a similar effect. There is also a "Casimir" torque, a twisting effect between birefringent plates in vacuum, which we once tried unsuccessfully to measure using a torsion pendulum in vacuum in my laboratory in Arizona. Birefringence is the strange property of certain crystals to support simultaneously two values of the refractive index, the effect discovered by Roemer's mentor Bartholin, which we encountered in Chapter 2.

Another issue which has only been understood since Einstein's 1905 paper is the way the speed of light sets a kind of measuring scale for the visible universe. It is possible that radiation from the most distant stars has not reached us yet, since the time when they were first created. The answer is connected with long-standing questions in astronomy concerned with the size of the universe and whether it is finite or infinite. Kepler believed it to be finite, with a limited number of stars. He pointed out that if it were infinite, you would see a star in every direction you looked—there would always be one at some distance away in any direction. Then one would expect the night sky to be bright, not dark, since the sky would be completely filled with stars. Why isn't it?

This issue of why the night sky is dark in certain directions has been discussed by astronomers down the centuries, from Halley and de Cheseaux to the writings of Wihelm Olbers in the nineteenth century, after whom it became known as *Olbers' paradox*. Building on earlier work, Edwin Hubble discovered in the 1920s that the universe was continuously expanding. He did this by measuring the Doppler shift, toward the red end of the spectrum, of light from stars at different distances from us. From the Doppler shift he could obtain the speed at which the stars were moving away from us. The measurements showed that the velocity at which galaxies are moving away from us is approximately proportional to their distance from us (Hubble's law). The Olbers paradox was explained by assuming that the red-shift of the light from the most distant receding galaxies would take the light beyond the end of the frequency scale for visible light. So the sky would be dark in directions in which those most distant stars were moving away so fast that their large red-shifts moved their light emission beyond the visible spectrum into the infra-red region.

This discovery that the universe is expanding is one of the greatest discoveries in human history. It must lead us to ask what happened at the beginning of the expansion. What would we see if we ran a movie of the expansion backwards? Hubble's discovery caused Einstein to modify his general theory of relativity to accommodate an expanding universe, originating with the big bang, when all matter was concentrated to a point. We now know that this occurred about 13.8 billion years ago. So that is the current date. Remember it! Everyone should know the date! Teach it to your children!

Hubble's law does not indicate that we on Earth are at the center of the universe, even though everything is moving away from us. The situation has been likened to raisins in a loaf of bread when it is rising due to the yeast. If the Earth were one of the raisins, all the other raisins would be moving away from it, even though it was not at the center. A simpler example is a rubber belt with studs on it. When you stretch it, all the studs get further apart from each other. In fact, it is space-time itself which is expanding.

The contemporary view of Olbers' paradox is that in directions where the night sky is dark, the light from the stars in that direction have simply not had time to get to Earth, since the big bang. We will therefore see a night sky which overall appears dark whether or not the universe is infinite in size. A light year is the distance light travels in one year. We only see the nearby stars, which, it might seem, are within a distance of 13.8 billion light years from us. This distance is the radius of a sphere, known as the cosmic event horizon, beyond which we cannot see anything. However, we must take account of the fact that the universe has expanded since the light was emitted, and when this correction is made, the radius of the visible universe becomes about forty-six billion light years. The number of galaxies (like our Milky Way) within this sphere is believed to be about two trillion. That limited number is not enough to fill the night sky with light. Accordingly the speed of light limits how far we can see.

The speed of light does set a measuring scale for the size of the visible universe, and was a crucial quantity needed to estimate the age of the universe. Two different methods give closely similar results for this age—extrapolation backwards from the measured expansion rate of the universe to the time of the big bang based on Hubble's law, and measurements taken from the cosmic ray background. These give an error of only about twenty million years (out of 13.8 billion). The age of the Earth, 4.5 billion years, is about a third of the age of the universe, and is related to the age of our Sun. It is believed that our solar system formed from the disk of material produced by the gravitational collapse of the molecular cloud which produced our Sun. The story of estimates of the age of the Earth is fascinating, and became critical to acceptance of Darwin's theory of evolution. It is another example of the way discoveries in science often result from imagining what might happen on a completely different scale of time, distance, speed, or energy. So I hope you will permit a brief digression on that fascinating topic.

The early estimates of the age of the Earth, based on either geological or biblical ideas, were far too short, varying from tens of thousands of years to about a hundred million. Kelvin produced the first physical calculation in 1862, based on Newton's law of cooling. He assumed that the Earth developed from a core of molten iron, and calculated how long it would take for the surface to cool down to our current temperature. His result, between twenty and a hundred million years, was later reduced to about twenty million years in around 1897. This was far too brief to allow for the evolution of the diverse forms of life we see on Earth, and so was used by critics of Darwin to debunk his theory. Geologists at the time, such as Charles Lyell, believed the Earth to be much older. At the time of his first estimates, Kelvin could not have known that the Earth's core is radioactive, greatly slowing down the cooling process. (Radioactivity was discovered in 1896.) Modern estimates of the age of the Earth (4.5 billion years) are based on measurements of the natural radioactivity of rocks, giving an age which does allow plenty of time for Darwin's evolution and origin of different species to develop. Life is thought to have started on Earth about 3.7 billion years ago.

In spite of many eloquent efforts to popularize it, Darwin's theory remains widely misunderstood. Many people think that evolution

occurs as a result of a conscious struggle for survival of the fittest for
their own betterment. But in its original and simplest form, the theories
of Darwin and Wallace seek to explain evolution by the action of pure
chance alone, acting over these very long time scales, which modern
physics gives us, but which many people still find very difficult to
accept.

The fable which is often told to exemplify the simplest form of
Darwin's theory imagines an island such as the Galapagos (which
Darwin visited) disrupted by an earthquake—a change in the environment—which creates two valleys, where only one existed previously.
Prior to the earthquake, a certain species of bird thrived, living on
worms they found in cracks in the rocks. After the earthquake, perhaps
the cracks in the rocks in the second valley were narrower and deeper.
Mutations in the genes in the DNA molecules of a bird would eventually produce a bird with a longer beak by chance, allowing it to obtain
more food in the second valley, and hence pass these genes on to more
offspring, just as our eye-color is determined by the genes of our parents and grandparents. In time, a very long time, the second valley
would be populated entirely by the more successful and better adapted
long-beaked birds who took over, producing more offspring because
they got more food. And they would eventually cease reproducing
with the short-beaked variety in the first valley, because they looked
different. In this way a new species of bird evolved—Darwin's
explanation for the origin of species by natural selection. (A species was
defined as plants or animals which reproduce together.) This theory led
to the idea than man and the apes shared a common ancestor, recently
confirmed by DNA testing. The important points of this fable are the
need for a new environment to create a new species, and the fact that
evolution occurs by chance rather than any act of will. The chance
mutations favorable to a new environment are said to improve the
organism's *fitness*. The modern view is that genetic mutations occur due
to copying errors when our DNA molecules are duplicated during cell
division. Evolution occurs because those birds which did not have
longer beaks simply died out in the second valley for lack of food. And
the beauty of the theory is that, because it is the result of chance mutations, it is extremely difficult to derive an argument which could show
how this process would *not* occur. It was crucial to note that acquired
characteristics, such as the strong arms of a blacksmith, could not be

inherited—the babies born to a blacksmith's family were not born with strong arms!

Of course the modern theory of evolution is far more complex than this, since these ideas had to reconciled with the statistical theory of mutations (classical genetics), epigenetics, and experimental confirmation. It was Gregor Mendel's 1865 studies of seven characteristics of pea plants (such as color and shape), which first revealed predictable patterns of inherited features. Darwin was unaware of Mendel's work—scholars debate whether he had a copy of Mendel's obscure paper on his shelves but had not read it. Recent research papers have claimed "direct observation of evolution." To observe evolution and the origin of a new species within the lifespan of the researcher requires a controlled change of environment and a very short lifespan for the plant or animal being studied, since it takes many generations for mutations to occur, most of which are not beneficial. One can imagine keeping a box full of flies (which normally live for about a month) in which their food is regularly supplied on the floor. How long (how many generations) would it be before we found that none of the flies in the box had any wings, which they don't need to get food or reproduce?

Evolution is now understood to operate at many levels, from that of the entire organism, down to the molecular level. But my broader point is that we often make advances in science by seeing things within a broader perspective, and in this case it was the scale of time itself, fixed partly by measurements of the speed of light, and leading to estimates of the age of the universe and our solar system, which gave credence to Darwin's theory.

Einstein went on to develop his *general theory of relativity* in 1916, in which he showed that gravity, which travels at the speed of light, bends light rays, and can also slow it down. The bending effect was confirmed experimentally by Sir Arthur Eddington and his team during a solar eclipse in 1919. (Eddington, in 1920, was the first to suggest that stars and our Sun are really continuously running hydrogen bombs, fusing hydrogen to produce helium in a nuclear reaction.) The tiny slowing-down effect has now also been measured. During the space probe *Viking*'s visit to Mars in 1979, a radio pulse was sent to and from the probe while it orbited Mars. A delay was measured due to gravitational fields, amounting to about two tenths of a millisecond and agreeing with the prediction of general relativity to within 0.2%.

This raises a fundamental question—how can the speed of light c be taken as a fundamental constant when it is affected by local gravitational fields? The strict answer is that it cannot, and Einstein himself commented that his special theory only works in the absence of gravity. Provided you avoid black holes, with their intense gravitational field, gravity normally has an extremely small effect on time. Since we now believe that most galaxies including our own Milky Way have a black hole at their center, it is remarkable that Einstein's general theory, which predicts the existence of black holes, also accounts for the large-scale structure of the universe.

Since 1905 there have been many claims for the transmission of information transfer at speeds exceeding that of light ("superluminal transport"). As we saw in Chapter 7, the phase velocity, which does not transfer information, may exceed c for light travelling in a medium, not in a vacuum. We've also seen one objection to the transfer of information faster than light—it may violate causality. This would allow things to happen before their cause, if there is not enough time for light to travel between the events. What happens if you murder your deceased great grandmother as a young girl by going back in time?

We could consider, for example, a plane wavefront of long but limited duration arriving at a large detector at some slightly inclined angle. Clearly the wavefield must arrive at one side of the detector before the other and the wavefront would indeed cross the detector at a speed greater than c. The paradox is resolved by the fact that no information can be extracted from this passage of the wavefront. Similarly, a small laser spot from Earth swept across the Moon cannot transfer information from one point on the Moon to another, even though the spot can move faster than light across the Moon, because the speed of the spot on the Moon is proportional to the distance to the Moon. The physicists Bilaniuk and Feinberg have proposed particles called tachyons, which, on theoretical grounds, could travel faster than light. However no experimental evidence for them has yet been found.

If we view two frames of reference approaching each other at high speed from a third, their closing speed can exceed c. However relativity is concerned with the analysis of one object in a moving frame as viewed from another. Quasars emit jets, detected by radio astronomy, which appear to move at lateral speeds exceeding c. But this has been found to be a relativistic projection effect, due to their motion being mainly directed toward the Earth.

Quantum mechanical tunneling is a process by which electrons can move through a barrier, when it is forbidden by the energy conservation rules of classical physics. It seems that virtual particles can tunnel faster than the speed of light, but again, no information can be sent this way. Many scientific papers have been written since 1905 about the puzzle of how long it takes electrons to "tunnel" through such a barrier. It was the subject of the Nobel Prize awarded to Leo Esaki in 1973, who invented the *tunnel diode*, which is capable of very fast switching. Tunneling experiments have also been undertaken using photons leaking across partially transparent mirrors and showing speeds in excess of the speed of light, but again no signals can be sent this way, as described in the book by Gerry and Bruno (2013).

Perhaps the most important and interesting case of possible superluminal information transfer to arise since 1905 comes from a paper which Einstein himself wrote with his colleagues Podolsky and Rosen in 1935, known ever after as "EPR." This paper has truly profound philosophical consequences for the ultimate nature of reality, so we need to go slowly to understand it. Much research, experimental and theoretical, has followed that paper, including the important "Bell's theorem," which connects the possibility of instantaneous action at a distance with the reality of quantum mechanical processes. Much of the theoretical development which arose from it lies behind the current development of quantum computers, and will surely soon earn a Nobel Prize. First we need to review some concepts from quantum mechanics.

To understand EPR and what it means for superluminal information transfer we first need to remind ourselves of Thomas Young's experiment. Then we can discuss the interpretation of quantum mechanical wavefunctions. Then we can move on to EPR, Bell's theorem, and more recent work.

In **Figure 5.1** we showed the actual diagram drawn up by Young in 1807 to explain his observation of interference between two beams of light, and commented that it could just as well describe ripples on a pond. Subatomic particles travel as a wave and arrive as a particle, as we have seen. The mystery of quantum mechanics emerges when we send in one photon (a quantum of energy) at a time, say once every minute. The wave from this one photon spreads out and illuminates both pinholes in Young's screen at points A and B in **Figure 5.1**. The wave continues from A and B to the far wall, but when it arrives, being a single photon, it makes only one dot on the wall. Over time, these "raindrops" build up a spatter of dots, like a Pointillist painting, which will eventually build up a pattern of vertical stripes. These interference fringes create the same pattern which a smooth continuous wave would have made, as for the ripples on a pond. Again, the photons have travelled as a wave but arrived as a particle. This one-at-a-time experiment was first undertaken using light by G.I. Taylor in 1909 at the suggestion of J.J.Thomson in Cambridge, using a gas flame as a source, smoked glass attenuators, and exposure times on his film of up to three months. He was able to wander off to watch a cricket match during the exposures!

Those are the experimental facts, and this experiment is now easy to do with modern equipment. It has also been done using a beam of electrons (which, unlike photons, have mass), as I've reviewed in my article on electron interferometry (Spence 2007). Diffraction of particle beams has recently been demonstrated with increasingly massive particles. These include the demonstration of diffraction by a beam of carbon buckyballs, which also have a wavelength, like light. The buckyball is a giant molecule shaped like a soccer ball, made of sixty or more carbon atoms. Quantum mechanics teaches that all bodies have a quantum mechanical wavelength. However there is a practical limit to this process of diffraction by massive particles. We are unlikely to see a beam of farm tractors split into separate beams of tractors after passing through a row of fine slits anytime soon, for many reasons. These include the process of *decoherence* which we will soon describe.

The theory of quantum mechanics seeks to explain this apparently impossible observation—how does the second photon know where to go to build up the right pattern of stripes if it has no memory of where the first one went? And is it possible to determine which slit a photon went through? These experimental results have generated the most intense debate among scientists for decades, resulting in many extraordinary theories about the nature of reality, including the *many worlds* theory, and many others. But don't forget the "shut up and calculate" advocates, who eschew philosophizing, and simply point out that if you follow the rules of conventional quantum mechanics, you get the right answer anyway, even if you don't understand why. This "Copenhagen interpretation" is a rule which works, but doesn't provide much understanding. It seems to presage the extremely popular "shut up and play" tutorials on *YouTube*, where you can learn to play jazz guitar much more quickly by watching the fingers and listening to the slow playing of a teacher who never speaks.

Most of post-Newtonian physics is concerned with calculating the likelihood that something will happen, just as we say "there is an 80% chance of rain tomorrow." Quantum mechanics calculates this 80% probability, unlike classical physics, which might claim to predict with certainty the exact motions of the planets, or the outcome from a snooker shot. Even then, there are inevitable problems with fluctuations in starting conditions and inevitable measurement errors. Darwin's theory of evolution and the theory of statistical mechanics are two other theories which deal in probabilities, not certainties. Thinking statistically comes naturally now to modern scientists, but it can be very difficult for the general public to do so. The report of a single fatal accident involving a driverless automobile tends to make people conclude that they are unsafe and should be banned. This occurs even if it were true that on average far fewer people per mile were killed by them than by ordinary cars with drivers, in which case driverless cars would actually be much safer.

Einstein never accepted the idea of a theory which made only statistical predictions—"I am convinced that God is not playing at dice" he famously said. It is the wavefunction, similar to that shown in **Figure 5.1** which gives this probability, and we need to understand the interpretation of this wave in terms of probability with a simple example.

We get this wavefield by solving a famous equation discovered by the Austrian Erwin Schrödinger in 1926. Schrödinger's equation is a wave equation, exactly as that used to describe waves on a pond or light waves, but extended into three dimensions and modified so that it can also describe particles which have mass, such as an electron. We have seen previously how to assign a wavelength to light. It was Louis de Broglie in 1924 who first showed that a wavelength could also be assigned to particles with mass, such as an electron, so that a beam of electrons could also make interference fringes.

In 1952 the German physicist Max Born won the Nobel Prize mainly for giving, in 1926, what has now become the standard interpretation of this wavefield. He proposed that the probability that a particle is detected at a particular place and time is proportional to the square of the wavefunction. The square of the wavefunction at a point, say midway between atoms, is proportional to the likelihood of detecting an electron at that place. His interpretation differed from the one

Schrödinger had initially proposed for his equation. For the electrons in atoms, for example, Schrödinger proposed that the square of the wavefunction was proportional to the density of electrons at each point. Both interpretations are used today.

Following his PhD in 1905 at Göttingen in Germany, among luminaries such as Hilbert and Minkowski, Born served as a radio operator in the First World War. As a Professor at Göttingen he presided over the greatest school of physicists in the twentieth century, including most of the founders of quantum mechanics. His list of pupils and associates reads like a list of all the great physicists of the twentieth century, including Oppenheimer, Fermi, Pauli, Delbruck, Teller, Wigner, and Heisenberg. Born was Jewish and emigrated to the UK when the Nazis rose to power in 1933, firstly to Cambridge, and then Edinburgh. He helped many friends and relatives to escape from Germany. In Edinburgh he was assisted by Klaus Fuchs, who later became an important spy for the Russians at Los Alamos on the Manhattan atomic bomb project. Born became a British citizen one day before the Second World War broke out, thus avoiding internment, and returned to Germany in 1952. Born died in 1970.

Born had worked closely with the younger Werner Heisenberg and Pascual Jordan to publish the first complete version of quantum mechanics, known as matrix mechanics, in 1925. He did this just before Schrödinger published his famous equation and method, known as wave mechanics, in 1926. An earlier version formulated by Bohr in 1913, now known as "the old quantum theory" had encountered difficulties. Bohr's theory, intended to explain the energies of spectroscopic lines, followed on from the first suggestion that energy was quantized into very small discrete units, by Planck in 1900. Einstein, in a second 1905 paper on the photoelectric effect, had proposed that light consisted of quanta, later called photons.

We can use Schrödinger's interpretation of his wavefield to interpret Young's pinhole experiment, as it has now been demonstrated using a coherent beam of electrons rather than laser light. Schrödinger treated the square of the wavefunction as a smeared-out map of the electronic charge density, as if all the charge concentrated in the particle had been smeared out all over the place around an atom or a beam. In that case Figure 5.1 would be a map of electron density, with maxima in the density at D and E, where the waves overlap and build up, and minima between them.

You may be thinking that all these abstract ideas about electron waves and the interpretation of the wavefunction cannot have much to do with our real word of tables and chairs, because quantum mechanics only appears to be relevant to the very smallest molecular scale. But in my work I've often used Schrödinger's interpretation of the electron wavefunction to measure and calculate maps of the electron density between atoms in solids. Since this electron density glues the atoms together by forming a chemical bond between atoms, without it our world of real materials would simply fall apart. Using electron diffraction, in a team led by my student, J.M. Zuo, we were able to make three-dimensional pictures of these covalent chemical bonds between atoms, as shown in our paper in *Nature* (1999).

These electronic bonds determine the arrangement of atoms in solids, and hence the form of matter in our universe. If the bonds between carbon atoms form in a particular way, connecting billions of atoms, we get the graphite in pencils or in coal. A different kind of chemical bond between atoms, due to a different electron wave interference pattern between them, produces a diamond. Like graphite, diamonds are made solely of carbon atoms, but with different properties from graphite, being hard rather than soft. In this way, this strange wavefunction, with the counterintuitive properties in Young's experiment, really does control the nature and properties of our real world, such as the hardness of diamond. The bonds between all atoms, including those between the carbon, oxygen, nitrogen, and hydrogen atoms which make up most of our bodies, are entirely quantum mechanical. The same process is responsible for the highly specific action of molecular drugs, which almost always act by binding to particular sites at locations between atoms in protein molecules in the body. This process was described by Paul Ehrlich, who won the Nobel Prize in 1908 and found a cure for syphilis, as "*Corpora non agunt nisi fixata*" or "no action without binding." The quantum mechanical interactions ensure that the drug sticks onto the correct place on the protein, otherwise, on the wrong site, it may cause side effects from the drug.

These quantum interference effects, such as those seen in Young's interference experiment, become less important as objects get bigger. As a wavefield enlarges within matter the particle it represents is more likely to be scattered inelastically, losing energy, perhaps to a phonon, a

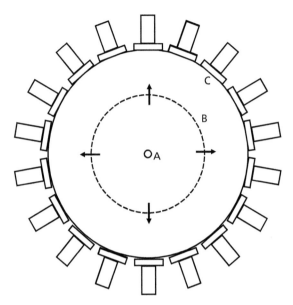

Figure 10.1 A spherical vacuum chamber surrounded by detectors (C). A snapshot of a spherical wavefront B is shown dashed, progressing from the small light source at A toward the detectors.

quantum of heat. This changes its wavelength and hence reduces its ability to interfere, as we discussed in connection with the beats between out-of-tune piano strings. This process is known as decoherence, and has been the subject of intensive research.

Let's explore an example of Born's interpretation of the wavefunction in quantum mechanics. **Figure 10.1** shows a point at the center of a hypothetical sphere emitting a spherical wave. The inner surface of the sphere is covered with small detectors. Creating a perfectly spherical wave is not so easy, since dipole radiators are normally used, as Hertz discovered, but devices which closely approximate a point source of radiation have recently been invented. An idealized point emitter would send out a spherical wavefront toward all the detectors, like dropping a stone in a still pond. Unlike the waves on a pond the energy of the particles represented by the electromagnetic wave occurs in discrete lumps or quanta, so we can send them out one at a time. What do we expect to happen at the detector? As for Young's experiment, we can use this spherical-wave solution of Schrödinger's equation to calculate the probability, according to Born, that any one of the detectors will record the arrival of a photon. For each detector, a photon either arrives or it does not, it cannot detect less energy than one photon. Since the amplitude of the wave is the same all around the inside of the sphere, the chances are equal for any of the detectors to record the arrival of a photon. As with the two-slit experiment, after the arrival of many photons we will see a spatter of photons distributed evenly over all the detectors, until it is eventually covered smoothly and uniformly. Such single-photon emitter devices have been made in many laboratories recently, and will no doubt soon be available commercially. You push a button and out pops one photon of light. It is said that the human eye, in a very dark room, when well adapted to the dark, can detect one photon.

It became popular to speak of the "collapse of the wavefunction" at one point on a detector. When a photon is detected, the wavefunction is then supposed to disappear everywhere else. It was argued that we cannot ask questions about the position of the photon *before* it is detected, *as though it does not exist at all until it is detected*. Philosophers have argued about whether *anything* exists before you look at it, or sense it in any way. What sense does it make to ask about a quantity before we make a measurement of it? The idea that the subatomic particles of which all matter consists do exist before they are detected is called *realism*, a concept which will be important in our discussion of Bell's theorem. But it could be that the act of observation causes the wavefunction to collapse—as Paul Davies has said, "Reality lies in the observations, not the electron." This, broadly, became known as the Copenhagen interpretation of quantum mechanics, championed by Niels Bohr, widely accepted today, but never accepted by Einstein. Until they are detected, particles must be described mathematically in all possible conditions, with various probabilities which change in time, but they cannot be said to exist in any of them prior to detection. As quoted in David Griffiths' textbook (1981), Pascual Jordan, in support of the Copenhagen interpretation, said "Observations not only disturb what is to be measured, they produce it...we compel the particle to assume a definite position."

Since Einstein's time, many other interpretations of the wavefunction have been developed. These include the *many worlds* interpretation, in which the wavefunction does not collapse, but generates an alternate history in an infinite number of alternative universes. Also proposed are the *consistent histories* approach, an *ensemble* interpretation, the *hidden variable* theory of de Broglie, Bohm, and Einstein (which led to Bell's theorem), and a *Bayesian* approach, amongst others. The Bayesian approach treats Born's probability function as simply updating our state of knowledge about an experiment, so that we could say that ultimately "it's all in the mind." Einstein's view (realism) was that the particle really was at the position at which it was detected just before it was detected. He believed therefore that quantum mechanics must be an incomplete theory, because it could only make statistical predictions, and must contain "hidden variables" which he hoped to expose. These hidden variables might provide the actual position of the particle at all times and places. An example of hidden variables from classical physics is the positions and momenta of all the molecules in a quantity n of gas. These are hidden in the simple gas laws relating temperature T, pressure P, and volume V. These simple laws, such as $PV = n R T$ (where R is a constant) don't contain these molecular positions and momenta, but depend on them.

There are many more theoretical interpretations of quantum mechanics, but that would take us too far from our theme—they are discussed routinely in journals, at conferences and in books on the foundations of quantum mechanics.

How wonderful it would have been to be a fly on the wall at the Baltimore conference dinner in 1884 when Kelvin and Rayleigh urged Michelson to try his experiment one last time! Apart from the famous fifth Solvay conference of 1927, where Einstein and Bohr debated quantum mechanics, my choice for the most exciting event to eavesdrop in physics in the twentieth century would have been Born's colloquium at Göttingen in 1926. The Austrian, Erwin Schrödinger, had been invited from Zurich, where he was Professor of Physics, into this lion's den to present his newer and alternative formulation of quantum mechanics. His approach became known as wave, rather than matrix, mechanics. Born and Heisenberg were in the audience, with pointed questions. Heisenberg never accepted Schrödinger's version of quantum mechanics, and likewise Schrödinger never accepted Born's interpretation of his wavefunction. "I don't like it, and I'm sorry I ever had anything to do with it" he once said. The difficult conceptual issue at the time was the statistical nature of electron transitions between different energy levels which can produce fluorescence. It was one thing to accept raindrops in Young's experiment, but even more difficult to accept that the new quantum mechanics could not predict exactly *when* an electron would jump from one energy level in an atom to another, emitting light. Similarly, the decay of individual atoms in a radioactive material like radium must be regarded as purely statistical. In these cases the theory makes only statistical predictions about the rate of transitions, not the exact time and date on which they will occur.

So in this dominant "Copenhagen" interpretation of quantum mechanics, due to Niels Bohr and Heisenberg, the idea is that while the particle is in its wave-like state, it exists in a kind of superposition of all possible states, only one of which will appear when it is detected. "States" mean things like its position, momentum, spin, and even energy. Each state is described by a wavefield such as that shown in **Figure 5.1**. Quantum mechanics predicts the likelihood that it will be detected in each of those states, but not which one it is in.

All this discussion applies to a single particle such as an electron, travelling either in a beam or perhaps bound to a proton to make a hydrogen atom. But quantum mechanics was soon expanded to include atoms which contain more than one electron, interacting with each other, such as helium with its two electrons orbiting a nucleus. Understanding quantum mechanics for systems with more than one electron is crucial for any understanding of the EPR paradox, entangled states, Bell's theorem, quantum computing, and the possible use of entangled states to communicate faster than light.

The first approach to the use of quantum mechanics for many-electron systems, the *self-consistent field approach*, was due to Douglas Hartree in 1927. He used iterations between two coupled Schrödinger equations, one for each electron, until they converged to give the separate wavefields for each of the electrons. Without electronic computers, these early calculations were done on mechanical calculators, which performed one numerical operation every time you pulled a handle, rather like a gambling machine. It was fortunate that Hartree's father was Professor of Engineering in Cambridge, so that Douglas had his father William's help with these machines. He devoted his early career to the application of numerical methods for solving differential equations using such machines. He built his own differential analyzer following a visit to MIT. He applied it to everything from computing railway timetables, to control theory, radio wave propagation, and fluid dynamics for aircraft flight. In 1939 he was able to publish with his father and Bertha Swirles the first *multiconfiguration Hartree–Fock* calculations, which had extended his method to include the intrinsic spin effects of electrons. The intrinsic spin of an electron means that each electron acts like a small bar magnet with its own quantized magnetic field. The Hartree–Fock method has since become perhaps the most popular algorithm used in molecular chemistry. It helps us to understand and predict chemical reactions, from the development of new drugs to the synthesis of new plastic materials and the catalysts used at oil refineries.

To understand quantum mechanics with more than one electron, we need to add one last additional principle—the *exchange interaction*. As we have seen, quantum mechanics requires us to consider every possibility in setting up our mathematical description of a system. For helium, we need wavefunctions for two electrons, not one, and these two electrons exert a force on each other. Both the force between the nucleus and each electron, and that between the electrons must be included in Schrödinger's wave equation, greatly complicating matters. Schrödinger's equation then predicts many wavefields over which the two electrons may be distributed, in many ways, depending on how the spins are assigned to the wavefield, known as orbitals. But these electrons are identical and fundamentally indistinguishable, so the complete wavefunction describing all of them must have the property that, if the coordinates of two electrons are interchanged, the square of the total wavefunction should not change. Then, for example, our charge density map would not change if the first and second electrons in a helium atom swapped positions.

To respect the anonymity of the electrons, a new fundamental principle (called "exchange") was developed. This does ensure that if two particles are interchanged, there will be no change in any measured quantity, and this determines the manner in which the wavefunction must be made up. The exchange energy is an amount of energy, equal to the difference between a correct calculation of energy in which the electrons are indistinguishable, and one in which they are not. This is the important energy which determines whether a material is magnetic, that is, if all the little bar-magnet electrons' spins line up with all their north poles pointing in the same direction. Alternatively, they may point in random directions, making a non-magnetic material.

The net result is that the total wavefield for a quantum system with many particles becomes a long summation of certain products of the wavefields for individual particles. We must then ask how Born's interpretation of wavefields applies to this wavefunction for many particles, the many-body wavefunction. It gives the probabilities for all the particles being found in particular states and positions.

But some many-particle states, appropriate to particular experiments, cannot be separated into a simple product of single-particle states, so that any measurement necessarily involves all the particles. These are called *entangled states*, and it is these that form the basis of the EPR paradox, and many other topics of current research, including quantum information theory, quantum cryptography (to prevent or enable hacking and the breaking of computer codes), quantum teleportation, and, as most but not all scientists believe, quantum computing. The early analysis of these entangled states first raised the prospect of superluminal information transport—communication at speed faster than light. This faster-than-light communication is greatly to be desired, since it would reduce the time from over an hour to communicate with *Cassini* at Saturn. It would also allow us to communicate with aliens in another galaxy and get a reply before we die. It would change our world forever, including fast trading on the stock market! You can find a summary of proposed superluminary schemes in Nick Herbert's book (1988). Given the current interest in quantum computers, the study of entangled states, on which the idea of faster-than-light communication depends, is a field of very active current research, all of which emerged in some sense from the 1935 EPR paper.

The idea of the EPR paper can best be described in Einstein's own words in 1933 at a Solvay conference:

> Leon Rosenfeld, one of Niels Bohr's collaborators, recalled that after a lecture by Bohr Einstein pointed out to Rosenfeld that if two particles interacted with one another and then flew apart, an observer who gets hold of one of the particles, far away from the region of interaction, and measures its momentum ... [will] from the conditions of the experiment, ... obviously be able to deduce the momentum of the other particle. If, however, he chooses to measure the position of the first particle, he will be able to tell where the other particle is ... is it not paradoxical? How can the final state of the second particle be influenced by a measurement performed on the first, after all physical connection has ceased between them?

The simplest practical example would be an atom which spits out a blue and a red photon in opposite directions at the same instant. However far away we are when we detect the blue photon, we *immediately* know that a person miles away will detect a red one, apparently getting this information faster than light could transmit a signal between us. In the early fifties, David Bohm proposed a version of this experiment, based on the decay of a pi meson (a sub-atomic particle) into an electron and a positron (a positively charged electron), which fly off in opposite directions when the pi meson decays. Since the meson initially has zero spin angular momentum, so must the total of the scattered particles, by the principle of conservation of momentum. We mentioned before that each electron acts like a tiny bar magnet with a north and south pole. The strength of this tiny magnet is proportional to the spin angular momentum, but it is the orientation of the north and south poles that we are interested in when we measure the spin of an electron. There are only two possibilities—the spin (north pole) may be upward directed, or downward. Before the positive and negatively charged electrons (bar magnets) zoomed off in opposite directions, their magnets cancelled out, lying in opposite directions. The result we measure for the direction of the spin depends on the orientation of the detector. By measuring the spin of the electron which went off to the left miles away, we would *immediately* know the spin of the positron which went to the right, since it must be the opposite, to preserve the original total of zero angular momentum (spin, or magnetic field) for both particles before they split up. This knowledge of the spin of the remote particle would be obtained instantly (action at a distance!) without even measuring the positron spin, as long as they remain in the same entangled state. *This suggests information transfer faster than light over arbitrarily large distances.* The particles are in a singlet configuration—an arrangement which ensures that they are indistinguishable. In the many experiments done since 1970 to test this idea, scientists have used the polarization of laser light in place of electron spin, allowing study of the two photons many miles apart, or, recently, between a satellite and the ground. The measurement on one side would collapse the wavefunction on both sides, to give a definite value for a measured quantity. The surprising results of these exciting experiments will now be explained, following some necessary background.

Einstein described this effect as "spooky action at a distance," and he did not believe it existed. He therefore came to believe that quantum mechanics was an incomplete theory, in which there were "hidden variables." Exposing or adding these, he believed, would create a theory which provided, simultaneously, the position and momentum of real particles, which existed before they were detected, a feature called *realism.* As we have seen, quantum mechanics cannot tell us anything about the state of particles before they are detected. The prevailing Copenhagen interpretation did not support realism. Einstein's additional insistence that information or causal influences could not be transferred faster than the speed of light, violating causality, was called *locality.* Locality means that separated events cannot influence one another instantaneously. This meant there could be no instantaneous "spooky action at a distance." The ability to instantly determine a quantum state far away from an interaction region is thus known as non-localism. A debate has followed Einstein's EPR paper, of ever-increasing intensity, as to whether the world and our reality is real (do things exist before you look at them) and local (no influences faster than light), or otherwise. The most widely accepted interpretation of quantum mechanics, the Copenhagen interpretation due to Bohr, says that it is neither real nor local. Most physicists would probably just say that in quantum mechanics you cannot ask questions about the state of particles before you detect them. And that causal influences cannot travel faster than light, which they apparently do in the EPR experiment.

The EPR paper also sought to show that quantum mechanics was inconsistent because remote measurements of positions and momenta could be shown to violate Heisenberg's uncertainty principle. The state of the remote particle was also determined without disturbing it, something which is not allowed in quantum mechanics, unless perhaps they all remained part of one system. Einstein firmly believed in the existence of a physical reality which was independent of the observer and causal. Notice that quantum mechanics in Bohm's experiment cannot predict which spin (up or down) will be detected on the left, only that the spin on the right must be the opposite—they are correlated (related). Note also, that the correlated spin results could only be discovered long after the measurement, when records of the measurements at both ends were compared. Classical physics, where this does not happen, is said to obey "local realism," since action (detection on one side) and effect (the state established on the other, far away) cannot be causally related unless there is time for light to travel between the two particles (localism), and, in Newton's view, particles also do exist before they are detected (realism). Newton was a fierce proponent of localism; he once wrote

That one body may act upon another at a distance through a vacuum without the mediation of anything else . . . is to me so great an absurdity that I believe no man who has in philosophical matters a competent faculty for thinking, can ever fall into

Many hidden variable theories have been developed since 1935. But correlations (relationships) between particle properties do seem to propagate instantaneously within entangled systems, regardless of their size. The crucial question is whether these instant correlations can be used to convey information faster than light.

In 1964, John Stewart Bell published a theoretical analysis of the EPR paper based on spin measurements on pairs of entangled electrons, which reached some very important conclusions ("Bell's theorem"). Bell's result was an *inequality* for EPR type experiments which can distinguish between the predictions of any hidden variable (real) theory which assumed locality, and those of quantum mechanics. In local theories, the outcome at one detector cannot affect the result from the other. The theorem thereby tests if local realism holds. Without getting into the details, based on measurements of the polarization of laser light beams sent out in opposite directions from a common excited atom, Bell's scheme could compare measured quantities A and B. If A was greater than B, then the world is real and local. But if B was greater than A, we could conclude that the world is not local and real, in which case the world is either not real, or not local, or neither. He made two assumptions to obtain his inequality—reality (that the particles have a well-defined spin before measurement) and secondly, locality (no faster-than-light communication between detectors). He then used the equations of quantum mechanics to show that B was greater than A, and hence that the local reality which Einstein favored is impossible. Local hidden variable theories satisfy Bell's theorem, while quantum mechanics does not. His theorem does not apply to non-local hidden variable theories. Bell himself produced a non-local hidden variable theory which resolved the EPR inconsistency. He showed that no local hidden variable theory could reproduce all the predictions of quantum mechanics, so that one cannot assume that the final spin-state information was simply given to real particles when they left the interaction region. His theorem shows that if quantum mechanics is correct (which it seems to be), then any hidden variable theory must be non-local. If you are interested in the details of Bell's theorem, I can recommend the article by David Mermin (1990), which derives it clearly in non-mathematical terms. The Bell inequality has also been gamified. On November 30, 2016 over 100,000 people on the web voted independently for different experimental detector arrangements to be applied to twelve different sets of apparatus on five continents, resulting again in support for quantum mechanics over a classical hidden variable theory and contradicting local realism. The results were published in 2018 by C. Abellan and colleagues (2018).

Experiments since have confirmed that Bell's inequality is violated, and so agree with the predictions of the Copenhagen interpretation, which supports a form of non-locality, and is non-real. Critics have pointed out possible loopholes in the early experiments, and the subject has become highly technical and somewhat philosophical, involving freedom of choice. The new experiments use laser beams and preserve their polarization over such very large distances that even at the speed of light there is time to change the experimental conditions while the photons are in flight. A famous "delayed choice" experiment by Alain Aspect and his group closed one important loophole, that during the flight of the particles the particles might somehow sense the orientation of the detectors. This loophole was closed by effectively setting those orientations at random very quickly after the particles had separated. How random was the choice, you might ask. If an individual person made the choice while the particles were in flight, you could ask whether that person really had free will. Their choice could have been predetermined in a fully quantum mechanical model of the universe which included them in the system. The detection of the two particles has also been made more rapidly than light could have travelled between the detectors, suggesting that causality is violated. In fact it is not, as we shall see.

If the two separated but entangled particles were travelling at the speed of light from the Sun, one going to a detector on Earth, you would have more than eight minutes to decide which component of spin you wanted to measure, assuming neither had any disturbing interactions on the way. The upshot of all these and more recent experiments is the finding that nature at the quantum level is certainly non-local, and a decline in popularity of Einstein's philosophy of realism (that particles exist physically before being detected, described by hidden variables). The Nobel laureate Tony Leggett has published an inequality which distinguishes between non-local hidden variable (realistic) theories which do not limit causality to the speed of light and quantum mechanics. Experimental tests based on this inequality have since falsified these non-local real theories. You can find a listing of simple experiments for undergraduates using lasers which test for local reality, EPR, and Bell's theorem at the website of Professor Mark Beck at Reed College (http://people.reed.edu/~beckm/QM/)

Unfortunately, it has since come to be understood that the fixing of a quantum system property (such as the polarization direction of a laser light beam at a satellite), which could be instantaneously determined by a measurement made a long way away (e.g. on Earth), cannot be used to transmit information. This is because the spin (or polarization) of the first measurement cannot itself be predicted. On Earth, there's always a fifty–fifty chance of measuring either polarization. Even if the measurements are repeated many times, the spooky action-at-a-distance effect gets smeared out in the overall statistics. Going back to Bohm's experiment with electrons and positrons, we cannot send binary bits between the electron detector on one side and that on the other detecting positrons miles away in the opposite direction to the electron travel, because we cannot predict the orientation of the spin we will detect at the electron side in the first place. So we cannot use non-locality for faster-than-light communications, and this has come to be known as the "no-signaling" theorem. Many of the related superluminal communications schemes which have been proposed (but not demonstrated) are summarized in the books by Tim Maudlin (2011) and Nick Herbert (1988). So the Copenhagen interpretation avoids violating causality.

We need to distinguish these experiments into the *foundations of quantum mechanics* using entangled states from the related experiments undertaken by Anton Zeilinger and others, which use entangled states in laser beams to transmit quantum-encoded messages; these are described in his book (2010). These experiments can create an identical remote copy of a quantum state, while destroying the original. The action is like a fax machine which shreds the original after reading it in, but transmits a faithful copy. The critical difficulty in these experiments is the need to avoid decoherence as the photons travel, such as would occur by scattering from molecules in the atmosphere. This "quantum teleportation" has been demonstrated by sending single laser photons between the Canary Island La Palma and the island of Tenerife, about eighty-eight miles away. And there are plans now by the Zeilinger group to send such messages from Earth to the geostationary satellite location, about 22,000 miles from Earth. Laser beams in space clearly avoid decoherence problems, however these problems return once we try to do something such as quantum computing with the teleported photons back on Earth.

The history of EPR, Bell's paper, and subsequent experiments has many twists and turns. It begins with Louis de Broglie, the scientist who won a Nobel Prize in 1927 for the idea in his PhD that if light has a wavelength, so too should electrons. And if there is a wavelength, there should be a wave equation—hence the Schrödinger equation. Later, de Broglie came up with the idea of a "pilot wave" to reconcile the wave–particle duality—this wave would guide real particles through the slits in Young's experiment. It became the first "hidden variable" theory. In 1932, the great Princeton mathematician John von Neumann, in his book on quantum mechanics, claimed to prove that hidden variable theories must be wrong. In this proof he made an error, picked up later by Hermann and Bell.

Einstein expressed his concern about the EPR problem to Rosenfeld in 1933, as quoted previously. Soon after, Einstein moved to Princeton, to work on the EPR paper with his assistant Rosen. Einstein was still learning English at that stage. The paper was written by Podolsky, who was thought to have the best English, although the title of the paper lacked the definite article. Einstein felt that since measurement on one particle confirmed the existence of the other, they must be real particles. He also believed that quantum mechanics must be incomplete—hence the need for additional "hidden" variables. There were response papers from Schrödinger and Bohr which did not resolve the issue.

After the Second World War, David Bohm, a member of the Young Communist League, completed his PhD in Berkeley under Oppenheimer, who moved to Los Alamos to run the Manhattan atomic bomb project, but was unable to take Bohm with him. Bohm stayed behind in Berkeley, from where he contributed theoretical work to the Manhattan project. Later, Bohm moved to Princeton and wrote his famous textbook on quantum theory, which was widely used in the following decades. I was taught from it in the 1960s. Einstein commented that it set out the Copenhagen interpretation about as well as it could be. It included the pi meson example of the EPR experiment. Soon after, Bohm published his own hidden variable theory, but he was hauled up before the McCarthy Un-American Activities committee because of his membership of the Communist League and his

Manhattan project work. He refused to testify, was charged with contempt of Congress, and suspended from Princeton. In 1951 he was acquitted, but with no job likely in the USA he moved to Brazil, then Israel, then, in 1961, to Birkbeck College London. He died in 1992.

John Bell (1928–90) came from a poor working class Northern Irish family. He moved from Belfast technical school to Queen's University in Belfast, graduating with first class honors in Physics in 1948. There he was taught by Peter Ewald, a founder of modern diffraction physics and crystallography. From Queen's he moved to a job at Harwell, the UK national laboratory, working on the European particle accelerator design for the CERN machine, which opened in Geneva in 1959. He obtained his PhD in 1956 and in 1960 moved permanently to CERN with his wife Mary, also an accelerator physicist. Bell was fascinated by Bohm's hidden variable theory with its pilot wave, and found time to study it and von Neumann's proof when he went on leave to the United States around 1963. He visited SLAC, the US Department of Energy accelerator near Stanford, and went on to Brandeis and Madison. During this leave-of-absence, with the uninterrupted time it provided him to think deeply, he wrote his two great papers. These showed that all hidden-variable theories are non-local, that the reality condition would require quantum mechanics to be non-local, and conversely that a locality condition excludes reality, so that local reality is excluded, and the universe is either non-local or not real. The result is commonly summarized as "no local reality." Both papers were largely ignored— one went to a journal which folded within a few years.

But Abner Shimony at Boston University read Bell's paper, and supervised a PhD student, Michael Horne, to work on the problem, which was completed in 1970. And in 1968, they learned of work on the same problem at Berkeley. They were surprised to see a paper submitted by John Clauser, then at Columbia, to the 1969 American Physical Society meeting. John is now retired, and a fellow sailing enthusiast at the Berkeley Yacht Club. Before long the three of them had met and published a founding paper in the field with Richard Holt, a graduate student at Harvard, describing possible experiments to test Bell's theorem, and they formulated a generalized Bell's theorem. Clauser subsequently moved to the Berkeley Physics department, in radio astronomy under Nobel laureate Charles Townes, inventor of the microwave laser. Townes left Clauser free to follow his EPR interests, with the result that the first experimental test of Bell's theorem was published by Clauser and Freedman in 1972. This concluded that Bell's theorem was violated, providing evidence against hidden-variable theories, and later that either realism or locality must be false. In the years since then, various loopholes with this experiment have been found and patched up.

To summarize these results, Einstein felt that the statistical nature of quantum mechanics meant that it was an incomplete theory because it could not make precise predictions, and hence that there were *hidden variables* which could be exposed to ensure reality and locality. But Bell's theorem showed that no hidden variable theory could replicate all the predictions of quantum mechanics in the Copenhagen interpretation. Subsequent experiments have shown that Bell's inequality is violated, supporting quantum mechanics. They exclude the possibility of local reality, that is, The world at the quantum level cannot be both real (where particles exist before they are detected) and local (where causal influences cannot travel faster than light). Leggett proposed a test for non-local realist theories as alternatives to quantum mechanics, which has been violated by experiments. This, and other work, resolves the issue of which we must give up, realism or localism, in favor of giving up both, and is consistent with quantum mechanics. The most common modern view among physicists is that a measurement on an entangled system of particles in a coherent state only has meaning for the entire system of particles, and measures a collective property.

What then is the status of the quantum mechanical wavefunction? When Bohr was asked about this after the 1935 paper, he said, "There is no quantum world. There is only an abstract quantum mechanical description." And it seems we must think of the two particles, however far apart, as components of a single (entangled) quantum state.

What should we conclude from all this, if we accept that the world is unreal and non-local, and so in agreement with the Copenhagen interpretation of quantum mechanics, which has made so many successful predictions? From a very large number of such predictions, the behavior of semiconductor devices and the "band-gap" in insulators has been among the most spectacular successes of quantum mechanics, giving us the modern computer and mobile phone. My own feeling is that the spooky non-local business is just another example of the way physics works. It is little different in principle from the electric fields we talked about earlier, going back to Faraday, and the jiggling of those electrons at the Goldstone dish in response to a tiny transmitter 746 million miles away. The many-body wavefunction, as interpreted by Born, performs exactly the same function as Faraday's and Maxwell's fields, giving us the correct mathematical description of a phenomenon we don't understand, in the sense that we don't know what the wavefunction represents physically. But perhaps that is just because I was taught quantum mechanics in the 1960s, when the Copenhagen interpretation was at the height of its popularity. And the idea we seem compelled to accept, that correlations in entangled systems are instantaneous but cannot convey information, can be compared with the earlier example of a laser beam sweeping across the Moon, or simply a shadow on a screen of a moving object.

As a working physicists I must confess that throughout my career I have always belonged to the "shut up and calculate" school, and I never saw the profundity of the ideas behind EPR and Bell's work until doing a lot of reading for this book. I now have "doubts"! For it seems clear that quantum mechanics cannot provide any information about a particle before it is detected, and it would require a hidden variable theory to do so. (Note that we could always make measurements at earlier times.) Like Newton's theory, I assumed that quantum mechanics was a work in progress which we should use until falsified by experimental results. But since the experiments confirming violation of Bell, and more recent inequalities, increasingly show that these hidden variable theories are unacceptable, it follows that quantum mechanics strongly suggests that reality (in the sense described by hidden variable theories) does not exist, and cannot be meaningfully described until it is observed.

When I run my hand across in front of a lamp directed at a distant wall, the magnified shadow image moves with much greater speed than my hand, by the ratio of the distance to the screen to that from the lamp to my hand. By taking the screen sufficiently far away, we can get any speed we like for the shadow, greater than the speed of light. But no information can be conveyed from one point on the screen to another by this method, nor can someone at one point on the screen use it to cause something to happen at another. So the experimental confirmation that Bell's inequality is violated showed that nature is fundamentally non-local, a shocking result at the time, which is still sinking in to the scientific community. In some sense it is a return to the instantaneous action-at-a-distance idea, but with this strange non-causal attribute.

Finally, in reading the literature about entangled states in our universe, it is worth remembering how very difficult it is to do these experiments on Earth, because complete coherence must be preserved for the entangled states, a situation that rarely occurs in our world of macroscopic objects. The results obtained so far do refer to a somewhat artificial situation. Except at the smallest scales, or perhaps in deep space, coherence is destroyed by the decoherence mechanism of inelastic scattering, which after a short time can result from room heating or any environmental influence. This decoherence is responsible for the transition to classical behavior in physics. The inelastic scattering changes the frequency of the waves, and so allows them to interfere only for a brief time, as we discussed earlier in connection with piano tuning and beat notes. But if we are prepared to consider the very shortest timescales, one could argue that entanglement remains in some form throughout the universe, so that Bohr's statement above about there being no quantum world could perhaps be inverted, to say that, if all timescales are considered, there is no classical world. For the Bell's inequality tests, it is not the spatial coherence which matters, but the preservation of a quantum state such as polarization or spin—the phenomenon of pulsar radiation reminds us that the polarization of radiation can be preserved over intergalactic distances. (Pulsars are rotating neutron stars which emit a lighthouse-like rotating beam of polarized radiation.)

In time-dependent quantum mechanics, the entire universe (which can have no external observers) could indeed be thought of as a single quantum mechanical state, however since it is changing in time, it cannot be in a single pure state (eigenstate). In this view, the measuring device or detector (including the brain of the person watching the detector) can all be included in one gigantic wavefunction. In the Copenhagen interpretation, however, it seems more efficient to impose decoherence and a transition to classical behavior at the detector. Decoherence may explain the transition to classical behavior at the detector, as vividly reviewed by Wojciech Zurek in his article, but it does not explain the way the quantum mechanical wavefunction collapses to give the value of unit probability at the place where the particle was actually detected, such as the photon in **Figure 10.1**.

While the Copenhagen interpretation has so far passed all tests with flying colors, it is generally accepted that this collapse of the wavefunction at the detector, or the "measurement problem" is its most unsatisfactory

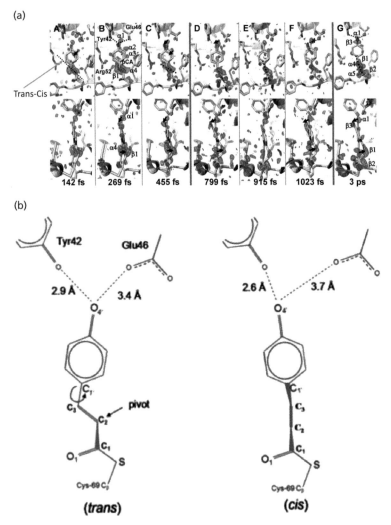

Figure 10.2 (a) The measurement problem of quantum mechanics. A protein (labelled "Trans-Cis") in photoactive yellow protein (PYP) wriggles as it detects (absorbs) a photon, as in the human eye responding to sunlight. These are X-ray movie frames running across the page in approximately 100 femtosecond increments, obtained using an X-ray laser. (b) The trans–cis isomerization reaction shown in more detail. A stick figure showing more clearly the wriggling motion of the trans–cis isomerization, before (left) and after (right) absorbing a photon. The long protein molecule rotates about the C_1–C_3 axis shown when it is hit by light. (From Pande et al. (2016).)

aspect, and there is a sense of unease that something is being missed. This was dramatically demonstrated in some of our own recent research in a collaboration led by Professor Marius Schmidt, shown in **Figure 10.2**. Here the recently invented X-ray laser has been used to obtain snapshots of a protein wriggling after absorbing (detecting) a photon. It's a vivid demonstration of Richard Feynman's comment, that when

> *An atom in the sun shakes; my eye electron shakes eight minutes later, because of a direct interaction.*

The wriggling protein forms part of the larger photoactive yellow protein, which helps certain bacteria find food by detecting light. This is the same process which occurs in the human eye when it responds to sunlight, so this is akin to the first event in human vision. The time resolution is about 100 femtoseconds (1 fs = 10^{-15} second) between movie frames running across the page, and the two rows of figures show orthogonal projections. The wriggling motion is known as a *cis–trans isomerization reaction*, and the motion, shown in more detail in Figure 10.2 (b), has the same effect on the molecule's shape as converting a left hand into a right hand. Getting such a movie is a complicated enterprise, as described in our paper. The proteins existed in micron-sized crystals, and were excited by laser light a certain time before the X-ray snapshots were recorded. This time was set to one frame time for the first frame, then twice that time for the second frame, and so on. At the same time we were collecting 120 snapshots, each with a certain delay between laser flash and X-ray pulse, per second. Then these Bragg scattering patterns had to be converted into maps of the electron charge density in the crystal. This was done for each different time delay between the optical excitation and the X-ray snapshot, one delay for each frame of the movie.

But the figure makes clear how arbitrary is the division between the point at which quantum mechanics applies (the incoming photon) and the "classical" detection event (the wriggling protein molecule). We could just as easily have continued a quantum mechanical description all the way from the protein up into the brain, where the sensation of sunlight is finally brought to consciousness, eight minutes after the motion of an atom in the Sun caused emission of a photon toward Planet Earth.

The events shown in **Figure 10.2** are the result of using two different radiations (light and X-rays) at different times. We could not get a snapshot of the protein at the precise instant that the photon arrived, because of experimental errors in the synchronization of the X-rays and laser "pump" light source, used to mimic sunlight. This error can now be brought down to perhaps 10 fs. So if we apply the uncertainly principle to these two different radiations applied at different times, we would find that the pump light was localized only to within a region whose diameter is about equal to the wavelength of the light. The X-ray image is similarly localized to a diameter of a few angstroms.

The issue of decoherence is the greatest problem which scientists trying to build quantum computers struggle with. The difficulty of addressing and reading out the entangled states in their quantum computer bits (qubits) is another major problem. You can learn more about these ideas from the very many books on the topic of quantum weirdness, such as those by John Gribbin (2014), and including the clear and accessible account by Tim Maudlin (2011), which also discusses superluminal schemes, or the more advanced treatment of Gerry and Bruno (2013).

In ending this book we might reflect on the persistent survival of the concept of instantaneous action at a distance, as we saw with Bell's theorem. The Aether itself, in its current form of the vacuum state, is anything but empty. In the theory of quantum electrodynamics, for example, the Higgs field, which gives mass to fundamental particles, becomes a new kind of Aether, assumed to fill the universe. The use of abstract mathematical concepts to describe the vacuum state, or Faraday and Maxwell's classical electric fields, or the wavefunction in quantum mechanics, cuts to the heart of what physics does. For the most part, physics attempts to derive predictive mathematical metaphors for an underlying reality which we don't understand. We should recall that mathematics itself does have its origins in the real world, as can be seen in the tortuous history of attempts to found simple arithmetic on logic, and in the importance and origins of geometry for mathematics.

Mathematical metaphors deepen our understanding of phenomena, give physical insight and allow us to predict phenomena—the simpler and more elegant they are, the closer we believe they are to the truth. This is partly so because simpler equations contain fewer adjustable parameters. As a French diplomat, commenting on a proposal by the British for a new colonial border in Africa is supposed to have said, "Yes, it works in practice, but does it work in theory?" But we cannot carry this principle, valuing elegance and symmetry above all else too far, as Sabine Hossenfelder points out in her book. The best theory must also provide a good fit to many experiments. I was taught that experiments test theory, not the other way around. The best you can do is to compare the predictions of competing theories with experimental measurements and their associated errors. If the error "bars" span both theories, one cannot choose between them.

Gravitational fields are another example of a field, similar to electric fields—physicists don't really know what they are—"they are what they do" perhaps? We do know how to calculate them and so make predictions, given some knowledge of their past or present behavior (known as boundary conditions). Similarly, our concept of energy is another example of an abstract quantity we cannot feel, see, or touch, but which is useful because simple mathematical expressions relate energy in all its forms. Maxwell himself saw even his equations for light propagation as metaphorical. According to Sir James Jeans in the 1931 Commemorative Volume on Maxwell's life, Maxwell wrote

> The analogy between light and the vibrations of an elastic medium . . . although its importance and fruitfulness cannot be overestimated, we must recollect that it is founded only on a resemblance in **form** between the laws of light and those of vibrations.

Einstein wrote in 1931 about how Maxwell has extended our idea of physical reality from the material particles of Newton's day, whose movements were governed by partial differential equations, to continuous fields governed by similar equations, not capable of any mechanical interpretation. But Einstein saw this transition, although a very great advance, as a temporary compromise which was logically incomplete. This was especially the case for quantum mechanics with its ghostly wavefunction inhabiting an unseen domain. But as the film director Woody Allen put it when asked about these invisible phenomena,

> There is no question that there is an unseen world. The problem is how far is it from midtown and how late is it open?

This process of devising a mathematical metaphor should be distinguished from *modelling*, which refers to the fitting of experimental data to a mathematical function defined by several parameters, by a process of trial and error. A good example of that might be the Wright brother's measurements of the lift force on their wing surfaces as a function of the angle of attack and speed of the airflow in their simple wind tunnel. They measured these, made graphs of the results, and engineers later found the best mathematical curve which fitted the points on these graphs. New tools based on neural network algorithms are now revolutionizing our ability to model systems through their ability to find these parameters quickly, and perform pattern recognition tasks such as face recognition efficiently. These include self-driving cars, search algorithms, and semiconductor fabrication plants. However, as many experts have pointed out, this doesn't help at all to understand the phenomena. The algorithms are prone to errors (they are "brittle"), with potentially very serious consequences for security applications (e.g. face recognition at airports). This is a particular problem with mathematical models of the brain, where neural network algorithms work well, but don't aid understanding. You cannot prove that they work, even if they usually do! It might be said that they are most useful when the cost of failure is low.

Stephen Hawking has spoken of "model-dependent reality," and we know that the human brain is constantly anticipating sensory input, modelling it in advance (perhaps also in our dreams). Perhaps our consciousness is nothing but the sum of all these models. And so we are left with a fundamental question. Is reality "out there" waiting to be discovered, or do we somehow impose our imagination on it to create it, a view favored by John von Neumann.

Maxwell, Newton, and Einstein were masters of the ability to give mathematical form to physical phenomena, based on their own physical insight and knowledge of mathematics. From this came the theory of light propagation at constant speed, and Einstein's abolition of the supporting medium, the Aether. In a book with Infeld, he pronounced an obituary for the Aether:

> The Aether revealed neither its mechanical construction nor absolute motion. Nothing remained of all its properties except that for which it was invented, i.e., its ability to transmit electromagnetic waves.

But why does this process of describing physical phenomena by mathematics work, and why does it predict the future, including planetary motions, chemical reactions, atom bombs, and computers. Why does it work so well in physics, unlike its application to other subjects such as economics, human behavior, the stock market, earthquake prediction, the spread of disease, or long-term weather forecasting? And are the resulting theories unique? Certainly the complexities of phenomena are important in determining their amenability to analysis, and we know from studies that having some kind of model or master equation is more important than having lots of data ("big data") for making predictions. An example is the case of successful short-term weather forecasting, which is based on the Navier–Stokes master equation. By contrast, collection of more data does not seem to improve economic forecasts or prediction of the stock market.

In an article referenced at the end of this book, the great physicist Eugene Wigner speaks of this "unreasonable effectiveness of mathematics in the natural sciences," of which Einstein and Maxwell's theories leading to the speed of light are paramount examples. Wigner points out that, with a different kind of imagination and curiosity, mathematicians are primarily interested in problems which advance the core agenda in their subject, mathematics, not physics. They therefore tackle problems of the greatest purely mathematical interest, unrelated to physical phenomena. Initially these problems are often of no practical relevance whatsoever, but often become so. Computer security codes and quantum computing are recent examples. Physicists instead use only the mathematical results which best describe their physical intuition—mathematics can never be a substitute for thought in physics. It seems fitting to end with Wigner's view on this, echoing a similar sentiment previously expressed by Kelvin—for most physicists the following is as close to "an act of faith" as they will ever admit to. Wigner writes

The miracle of the appropriateness of the language of mathematics for the formulation of the laws of physics is a wonderful gift which we neither understand nor deserve. We should be grateful for it and hope that it will remain valid in future research and that it will extend, for better or for worse, to our pleasure, even though perhaps also to our bafflement, to wide branches of learning.

APPENDIX I

Numerical Values

Quantity	Symbol	Modern value	Units
Speed of light	c	299,792,458 exact	m/s
Earth's speed around Sun		30 (67,000 mph)	km/s
Radius of the Earth	a	6,378,140 (3963 miles)	m
Speed of Sun in Milky Way		230,000	m/s
Trip by Sun around Milky Way		225 million	years
From Sun to Earth (1 AU)	AU	149,597,870.7 (93 million miles)	km
Sun/Earth travel time for light	τ	499	s
Length of year	T	365.256 x 86,400	s
Solar parallax	δ	8.79414	arc sec
Annual aberration	θ	20.4955	arc sec
Earth orbit eccentricity	e	0.0167	
Earth–Moon distance		238,900	miles
Speed of Earth's surface due to rotation of Earth at equator		1000	mph

The following relations hold among these quantities:

$AU = c\,\tau$ and $\sin \delta = a/AU$ and $\theta = (360 \times 3600/T)\,\tau\,/(\sqrt{(1-e^2)})$

Roemer's Report of his Measurement of the Time Taken by Light to Travel a Distance Equal to the Earth's Diameter

Translated into English for publication in *Philosophical Transactions*, Vol 12, June 25, 1677, page 893. This figure referred to by Roemer is figure 2.2 in this book.

(893)

A Demonstration concerning the Motion of Light, *communicated from* Paris, *in the* Journal des Scavans, *and here made English.*

PHilofophers have been labouring for many years to decide by fome Experience, whether the action of Light be conveyed in an inftance to diftant places, or whether it requireth time. M. *Romer* of the R. *Academy* of the Sciences hath devifed a way, taken from the Obfervations of the firft Satellit of *Jupiter*, by which he demonftrates, that for the diftance of about 3000 leagues, fuch as is very near the bignefs of the Diameter of the Earth, Light needs not one fecond of time.

Let (in *Fig.* 11.) A be the *Sun*, B *Jupiter*, C the firft Satellit of *Jupiter*, which enters into the fhadow of *Jupiter*, to come out of it at D ; and let EFGHKL be the *Earth* placed at divers diftances from *Jupiter*.

Now, fuppofe the *Earth*, being in L towards the fecond Quadrature of *Jupiter*, hath feen the firft Satellit at the time of its emerfion or iffuing out of the fhadow in D ; and that about 42½ hours after, (*viz.* after one revolution of this Satellit,) the *Earth* being in K, do fee it returned in D ; it is manifeft, that if the Light require time to traverfe the interval LK, the Satellit will be feen returned later in D, than it would have been if the Earth had remained in L, fo that the revolution of this Satellit being thus obferved by the Emerfions, will be retarded by fo much time, as the Light fhall have taken in paffing from L to K, and that, on the contrary, in the other Quadrature FG, where the *Earth* by approaching goes to meet the Light, the revolutions of the Immerfions will appear to be fhortned by fo much, as thofe of the Emerfions had appeared to be lengthned. And becaufe in 42½ hours, which this Satellit very near takes to make one revolution, the diftance between the *Earth* and *Jupiter* in both the Quadratures varies at leaft 210 Diameters of the *Earth*, it follows, that if for the account of every Diameter of the *Earth* there were required a fecond of time, the Light would take 3½ minutes for each of the intervals GF, KL ; which would caufe near half a quarter of an hour between two revolutions of the firft Satellit , one obferved in FG, and the other in KL, whereas there is not obferved any fenfible difference.

Yet.

(894)

Yet doth it not follow hence, that Light demands no time. For, after M. *Romer* had examin'd the thing more nearly, he found, that what was not senfible in two revolntions, became very confiderable in many being taken together, and that, for example, forty revolutions obferved on the fide F, might be fenfibly fhorter, than forty others obferved in any place of the Zodiack where *Jupiter* may be met with; and that in propor- tion of twenty two for the who e interval of H E, which is the double of the interval that is from hence to the Sun.

The neceffity of this new Equation of the retardment of Light, is eftablifhed bv all the obfervations that have been made in the *R. Academy*, and in the *Obfervatory*, for the fpace of eight years, and it hath been lately confirmed by the Emerfion of the firft Satellit obferved at *Paris* the 9th of *Aovember* laft at 5 a Clock, 35'. 45". at Night, 10 minutes later than it was to be expected, by deducing it from thofe that had been obferved in the Month of *Auguft*, when the *Earth* was much nearer to *Jupi- ter* : Which M. *Romer* had predicted to the faid Academy from the beginning of *September*.

But to remove all doubt, that this inequality is caufed by the retardment of the Light, he demonftrates, that it cannot come from any excentricity, or any other caufe of thofe that are commonly alledged to explicate the irregularities of the *Moon* and the other Planets ; though he be well aware, that the firft Satellit of *Jupiter* was excentrick, and that, befides, his revo- lutions were advanced or retarded according as *Jupiter* did approach to or recede from the Sun, as alfo that the revoluti- ons of the *primum mobile* were unequal ; yet faith he,thefe three laft caufes of inequality do not hinder the firft from being mani- feft.

APPENDIX 3

Foucault's Spinning Mirrors

In Figure A3.1 the light source S_1, off the page to the right of the diagram, (and previously shown in Figure 5.6) is imaged onto the surface of a spherical mirror M2 at S′, after reflection by the rotating mirror M1. The distance D between the rotating mirror and the spherical mirror is just equal to the radius of curvature of the mirror. If M1 were stationary, and we stood at S′, we would see our reflection, a virtual image behind the mirror, at S″. So the lens L forms an image of this virtual source S″ in the eye of the observer off to the right. The important point, and this was one of Focault's main innovations, is that, as shown in Figure A3.2, if we slowly rotate the mirror a little, because of the changes in angles shown (which preserves the specular mirror-reflection condition), although the mirror rotates, the position of the virtual image S″ does not. So the bright spot to the right of the focus of lens L in figure A3.1 would not move sideways if the mirror were slowly rotated. This works, provided the spot S′ remains within the short arc of the mirror M2, so that it is reflected. The flickering of the spot due to this effect (as the mirror rotates through 360°)

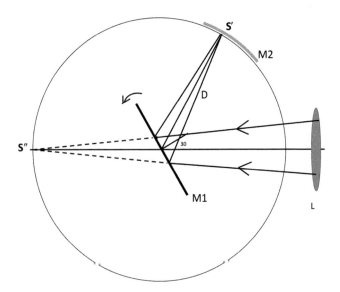

Figure A3.1 Focault's mirrors.

goes away if the spot is produced more frequently than the human eye can respond to, just as we see continuous motion in a movie if the frames are presented more frequently than about thirty times every second. Foucault reduced it further by making both sides of the mirror reflective. Foucault had solved Arago's synchronization problem by providing a reference position for reflected light at low rotation speeds (when it flows continuously), which did not depend on the orientation of the mirror.

Figure A3.2 shows the situation for a slowly rotating mirror. The situation is quite different if the mirror M1 rotates rapidly (at about 500 revolutions per second in Foucault's experiment). Now the light reflected from the curved mirror is chopped up into pulses, and the light pulse returning from the curved mirror finds that it has changed orientation while the light was travelling to S′ and back (about twenty meters in Foucault's experiments for the round trip). Then, as shown in Figure A3.3, the rotating mirror reflects an image of S′ from a different direction, from a virtual source at S″. The light arrives with the mirror M1 in the grey position, but is reflected from it in the black position. Then the bright spot seen by the experimenter to the far right will appear to move sideways (by about 0.7 mm in Foucault's experiment), by an amount proportional to the angle the mirror has moved through while light travelled the distance 2D. So one could expect the distance x in Figure 5.6

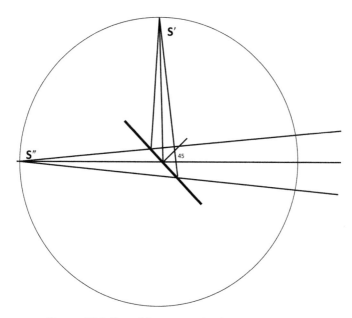

Figure A3.2 Focault's mirrors: slowly rotating mirror.

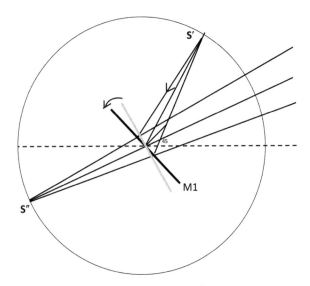

Figure A3.3 Focault's mirrors: rapidly rotating mirror.

to increase as the speed of the rotating mirror was increased, deflecting the returning beam through a larger angle as the mirror turns through a larger angle while light travels to M2 and back to M1.The speed of light can then be obtained by making a graph of the deflection of the spot against the speed of the mirror rotation. The slope of this graph will give the speed of light in terms of known quantities.

Einstein's Theory of Relativity, Simplified

Figure A4.1 shows our rail car now fitted with a light detector on the floor directly below the lamp, which lies a distance z above the detector. We aim to find the time ΔT_t it takes light to reach this detector, as measured by an observer in the train, and compare that with an estimate of the time ΔT_o made by someone on the ground. For the observer on the train, the time is just $\Delta T_t = z/c$, with c the speed of light. For an observer on the ground, the train is seen to move from A to B, a distance $v\Delta T_o$, while the light is travelling down to the floor, so the only light which can reach the new displaced detector position at C must travel along the path OC, to hit C in the middle of the car on the floor. Then by Pythagoras's theorem, the light must travel a distance $L = \sqrt{(z^2 + (v\,\Delta T_o)^2)}$, so the time taken by the light is $\Delta T_o = L/c$. After some algebraic manipulation of these three equations we obtain

$$\Delta T_o = \gamma\,\Delta T_t ,$$

where $\gamma = 1/(\sqrt{(1 - v^2/c^2)})$ as defined previously. The faster the train goes, the larger is γ, meaning that more seconds ΔT_o will tick on the land-based clock for a given number of seconds on the clock on the train, which will therefore be seen to run slow when viewed from the ground. Hence "moving clocks run slow." The famous twin paradox considers a pair of twins, one of whom sets off in a rocket at high speed, eventually returning home to find her twin brother much older! This is correct, and the situation is not symmetrical, because the travelling twin must experience acceleration to return home (violating the conditions of special relativity), unlike her brother. So we cannot easily analyze the situation symmetrically from the point of view of the travelling twin.

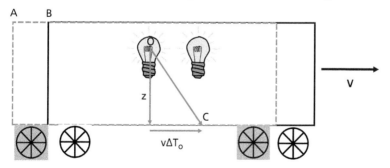

Figure A4.1 Time dilation in special relativity. The distance from O to C is L.

The final bizarre result of special relativity ("moving sticks get shorter") can be obtained by considering light from a lamp emitted at one end inside the railway car and reflected back to the same wall. It is then found that the dimensions of objects are contracted in.the direction of their motion by an amount γ when measured from a "stationary" frame. Taken together, these time and length changes form the key elements of the Lorentz transformation which convert estimates of time and length made in one moving frame to those made in another.

How to Measure the Speed of Light with a Microwave Oven and Pizza Dough

Microwave ovens were developed from the cavity magnetron oscillator developed for radar in the Second World War. They generate an intense beam of gigahertz radio waves, which are strongly absorbed by the water and fat in food, causing molecules with a dipole moment to rotate, and thus heat them. Understanding the absorption of radar beams by water in the atmosphere, limiting the range of radar, was one of the great stories of scientific endeavors of the last century, leading to the invention of the laser, again based on some equations of Einstein published in 1916.

There are three kinds of waves. You can see all of these on a jump-rope. They are pulses, running waves, and standing waves. A running wave is a normal ocean wave with a single wavelength, where all the crests move forward together. A pulse is a localized peak which runs along—what you see if you whip the end of the rope once quickly—a pulse runs along it. More complicated is the standing wave, which is generated on the strings of a guitar or piano, and in your microwave oven. It occurs if the ends of the vibrating medium (such as a guitar string) are fixed. It's shown in Figure 9.4. It is a sinusoidal disturbance, just like a running wave, with the important difference, that, at any point, the motion rises and falls to the same height every time. But each neighboring point rises and falls to a different height. The rising and falling crests don't move forward. The wave height is slightly different (making a wave-shaped envelope) as you move along the wave. For some points, called nodes, the height is always zero (this is how guitar tuning using harmonics works, by lightly touching a string part-way along its length, imposing a node). So at those nodes, the rope is stationary all the time.

Modern microwave ovens create a standing wave at about 2.45 GHz frequency, in the same range as a mobile phone, or the communications with *Cassini*, with a wavelength of 12.2 cm (4.8 inches). The standing waves could be a problem, because the places on the pizza where this radio standing wave has a node will not get hot—at those places there is no radio wave energy. To address this, the manufacturers install the rotating dish, which you put the food on.

To measure the speed of light, we first measure the wavelength. This is best done by removing the rotating dish and placing some smoothed-out flat pizza dough in the oven, and running it briefly at a low setting. The radio standing wave will impress a pattern of stripes on the dough (not cooked at the nodes,

maximum heat at the anti-nodes halfway between the nodes). You just measure the spacing between the strips, and this will be half the wavelength.

Now we know that the speed of light is equal to the frequency multiplied
by the wavelength. The frequency of the magnetron is usually written on the
back of the oven. If our stripes were 6.1 cm apart, giving wavelength 12.2 cm,
then the speed of light would be the frequency divided by the wavelength, or
2,450,000,000 x 12.2 cm = 294,000,000 meters per second. You can compare your
number with the values given in Appendix 1 and others given in Chapter 6 on
Maxwell.

A simpler way to measure the speed of light can be undertaken with any
modern oscilloscope. Even an oscilloscope of modest cost can measure frequencies up to 100 MHz. If you send a very short pulse from a pulse generator
down a coiled-up wire which is a hundred meters long, it will take about
100/300,000,000 = 0.3 microseconds to get to the end (connected to the oscilloscope), a time which is easily resolved by the oscilloscope. We can set up a two-
channel oscilloscope to display the pulse when it leaves the pulse generator
and enters the long wire, and, on the second channel, when it gets back from
the end of the wire. The return pulse will be seen to be delayed with respect to
the pulse sent out. Knowing the length of the cable and the time taken by the
pulse, you can calculate the "speed of electricity," as Wheatstone tried to do.
As for the Atlantic cable, the pulse will be distorted on its return unless you
use good quality coaxial cable and pay attention to the impedance matching at
either end.

Sources and References

Firstly I must thank Archie Howie in Cambridge UK and Kevin Schmidt in Arizona, who have been my mentors in theoretical physics for many years. I also wish to thank Professor James Goding particularly, for his detailed suggestions and many improvements to the text in style and content, and to Drs Joe Chen and Andrew Spence for their corrections. The community of academic historians of science can provide far more detailed information on each individual topic in this book. A good listing of many of these (and to the hundreds of densely mathematical nineteenth-century papers) can be found in the preface to Darrigol's (2000) superb book on electrodynamics in the nineteenth century, in Whittaker's (1910) magisterial volumes, and in Bruce Hunt's (1994) book. For physics students, I can strongly recommend Longair's (2003) book for his outline of the rich background of historical ideas which have contributed to theoretical physics, all in consistent modern notation. Otherwise, the following books and papers will provide more detailed background on aspects of this remarkable intellectual adventure, many at about the level of this book.

Abellán, C., Acin, A., Alarcón, A., Alibart, O., Andersen, C.K., Andreoli, F., Beckert, A., Beduini, F.A., Bendersky, A., Bentivegna, M. and Bierhorst, P., 2018. "Challenging local realism with human choices". *arXiv preprint arXiv:1805.04431.*

Alvager, T., Farley, F., and Kjellman, J. (1964). "Test of the second postulate of special relativity in the GeV region." *Phys. Lett.* 12, 260.

Aughton, P. (2004). *The Transit of Venus: The Brief Brilliant Life of Jeremiah Horrocks, Father of British Astronomy.* Weidenfeld and Nicolson. London.

Barrow, J.D. (2002). *The Constants of Nature.* Vintage, Random House. New York.

Beaglehole, J.C. (1968). *The Journals of Captain Cook.* Cambridge University Press, Cambridge.

Bork, A.M. (1963). "Maxwell, displacement current, and symmetry." *American Journal of Physics* 31, 854.

Burton H.E. (1945). "The optics of Euclid". *Journal of the Optical Society of America* 35, 357.

Campbell, L. and Garnett, W. (1884). *The Life of James Clerk Maxwell.* Macmillan. London.

Cohen, I. Bernard (1942). *Roemer and the First Determination of the Velocity of Light.* The Burndy Library Inc. New York.

Crewe, H. Ed. (1981). *The Wave Theory of Light.* Memoirs of Huygens, Young and Fresnel. Arno Press, New York.

Darrigol, O. (2000). *Electrodynamics from Ampere to Einstein*. Oxford University Press. Oxford UK. A comprehensive, modern historical view, providing depth and insight. Equations in all four systems of units, and relationship between them.

Darrigol O. (2012). *A History of Optics from Greek Antiquity to the Nineteenth Century*. Oxford University Press, Oxford UK.

Dawkins, R. (2016). *The Selfish Gene*. Oxford University Press, Oxford UK.

Dorsey, N.E. (1944). "The velocity of light." *Trans. Am. Phil. Soc.* 34, 1–110. Comprehensive technical review, particularly in analysis of errors, from Fizeau to 1940.

Einstein, A. (1905). "On the electrodynamics of moving bodies." *Annalen der Physik* 17, no. **891**, 50.

Einstein, A., Podolsky, N., and Rosen, B. (1935). "Can quantum-mechanical description of physical reality be considered complete?" *Phys Rev.* **47**, 777–80.

Einstein, A. and Infeld L. (1938). *The Evolution of Physics*. Cambridge University Press, Cambridge.

Faccio, D. and Clerici, M. (2006). "Revisiting the 1888 Hertz experiment" *Am J. Phys.* 74, 992.

Fahie, J.J. (1899). A *history of wireless telegraphy 1838–1899*. Blackwood , London.

Filonovich, S.R. (1986). *The Greatest Speed*. MIR (Moscow). Excellent short account of the topic of this book, with simple equations.

FitzGerald, G. F. (1883). "On a Method of Producing Electromagnetic Disturbances of Comparatively Short Wavelengths". In *Report on the 53rd Meeting of the British Association for the advancement of Science*. (1884) John Murray. London. p. 405.

Frercks J. (2005). "Fizeau's research program on Aether drag." *Physics in Perspective* 7, 35–65.

Galilieo Galilei (1632). *Dialog Concerning the Two Chief World Systems*. Ed. S.J. Gould. The Modern Library. Random House New York. (2001).

Ganci, S. (2013). "Was Poisson's spot an intentional discovery?" *Optik.* 124, 3906.

Gerry, C. and Bruno, K. (2013) *The Quantum Divide: Why Schroedinger's Cat is Either Dead or Alive*. Oxford University Press, Oxford UK.

Gribbin, J. (2014). *Computing with Quantum Cats: From Colossus to Qubits*. Prometheus Books. Kindle Edition.

Griffiths, D.J. (1981). *Introduction to Electrodynamics*. Prentice-Hall, New York. Excellent modern undergraduate text, widely adopted.

Heaviside, O. (1893). *Electromagnetic Theory*. The Electrician Printing and Publishing Co. London.

Herbert, N. (1988). *Faster than Light: Superluminal Loopholes in Physics*. Dutton, New York.

Hertz, H. (1893). *Electric Waves*. Dover Edition. New York.

Hertz, Johanna (1977). *Heinrich Hertz: Memoirs, Letters, Diaries*. 2nd Edition. San Francisco Press. San Francisco.

Hirschfeld, A.W. (2001). *Parallax*. Henry Holt, New York.

Holmes, Richard (2008). *The Age of Wonder. How the Romantic Generation Discovered the Beauty and Terror of Science*. Harper Collins, New York.

Horrebow, Peder (1735). *Basis Astronomiae*. Public Domain.

Hossenfelder, Sabine (2018). *Lost in Math: How Beauty Leads Physics Astray*. Basic Books, New York.

Hughes, Ivor (2010). *Before We Went Wireless*. Images from the past, Bennington, Vermont. The story of David Hughes's life.

Hunt, B. (1994). *The Maxwellians*. Cornell University Press. Excellent account of those who came after Maxwell (Lodge, FitzGerald, Heaviside, Hertz, Larmor) and their contributions as founders of modern classical electrodynamics.

James Clerk Maxwell (1931). A Commemorative Volume 1831–1931. Cambridge University Press. Cambridge UK.

Jones, Bence (1870). *Faraday's Life and Letters,* Volume II. Longmans, London.

Kumar, Manjit. (2008) *Quantum: Einstein, Bohr and the Great Debate About the Nature of Reality*. Icon Books. Kindle Edition.

Larmor, J. (1900). *Aether and Matter*. Cambridge University Press, Cambridge.

Levitt, T. (2013). *A Short Bright Flash*. Augustin Fresnel and the birth of the modern lighthouse. Norton, New York.

Lodge, O.J. (1889). *Modern Views of Electricity*. MacMillan and Co., London.

Lodge, O. J. 1892. "On the present state of our knowledge of the connection between ether and matter: an historical summary". *Nature*, **46**, 164–165.

Longair, M. (2003). *Theoretical Concepts in Physics*. Cambridge University Press, Cambridge. Superb account of the history of fundamental theoretical concepts in physics, together with all equations in consistent modern form.

Lorentz, H. A. (1895) "Attempt of a theory of electrical and optical phenomena in moving bodies". *Leiden: EJ Brill, Leiden April 1895*.

Marciano J.B. (2014) *Whatever Happened to the Metric System*. Bloomsbury, New York.

Maudlin, Tim (2011). *Quantum Non-locality and Relativity*. Wiley-Blackwell. New York.

Maxwell, J.C. (1870). *Address to the mathematical and physical section of the British Association*. in: Fortieth Meeting of the British Association for the Advacement of Science, (1871). John Murray. London. p. 1.

Maxwell, J.C. (1873) *A treatise on electricty and magnetism*. Oxford. Clarendon Press.

Maxwell, J.C. (1889) in Encyclopedia Britannica, 9th edition, London. Volume 8. *The Aether*

Mermin, D.N. (1990). *Boojums All The Way*. Cambridge University Press, Cambridge.

Michelson, A.A. (1903). *Light Waves and their Uses*. University of Chicago Press, Chicago.

Newton, I. (1704). *Opticks*. Dover publications (1952). New York.

Pais, A. (1982). *Subtle is the Lord.* Oxford University Press, Oxford. The best biography of Einstein, containing much historical and technical information from someone who worked with him.

Pande, K. et al. (2016). "Femtosecond structural dynamics drives the trans/cis isomerization in photoactive yellow protein." *Science*, **352**, 6286.

Preston, Thomas (1928). *The Theory of Light.* Thomas Preston. Fifth edition. Macmillan, London.

Rhodes, R. (1986). *The Making of the Atom Bomb.* Simon and Schuster, New York.

Robinson, A. (2006). *The Last Man who Knew Everything.* Pi Press, New York.

Shankland, R.S. (1964). "The Michelson-Morley experiment." *American Journal of Physics*, **32**, 16.

Shaw, J.A. (2013). "Radiometery and the Friis transmission equation." *Am. J. Phys.* 81, 33.

Simpson, T.K. (2006). *Maxwell on the Electromagnetic Field.* Rutgers University Press. New Brunswick. Contains his three great papers on electrodynamics and detailed analysis.

Spence, J.C.H. (2007). "Electron Interferometry" in *Compendium of Quantum Mechanics.* p. 188. Ed F. Weinert. Springer. New York. See also Spence, J.C.H. (2013). *High resolution electron microscopy.* 4th Ed. Oxford University Press. New York.

Stewart, A.B. (1964). "The discovery of stellar aberration." *Scientific American*, 210, 100.

Susskind, C. (1964). "Observations of electromagnetic radiation before Hertz." *Isis* **55**, 32–42.

Thompson, S.P. (1901). *Michael Faraday, his Life and Work.* Cassell and Co., London.

Tobin W. (1993)."Toothed wheels and rotating mirrors." *Vistas in Astronomy* **36**, 253–94. See also, *The Life and Science of Leon Foucault* by W. Tobin. Cambridge University Press, (2003).

Velten, A. et al. (2013). "Femto-photography: capturing and visualizing the propagation of light." *ACM Transactions on Graphics (ToG)* 32 no. 4 p. 44.Velten

Webb, J. (2003). "Are the laws of nature changing with time?" *Physics World* April 2003, p. 33.

Weinberg, S. (2015). *To Explain the World.* Harper. New York. Excellent survey of the history of astronomy from the Greeks to the time of Newton, with simple mathematical derivations in appendices.

Whittaker, E.T. (1910) (and later editions with important changes). *A History of the Theories of Aether and Electricity: From the Age of Descartes to the Close of the Nineteenth Century.* Longmans, Green, London. Comprehensive, advanced and authoritative, with full mathematical analysis in modern notation. British emphasis

Wigner, E. (1960). "The unreasonable effectiveness of mathematics in the natural sciences," in *Communications in Pure and Applied Mathematics* 13(I), (February 1960). John Wiley & Sons, New York.

Wulf, Andrea (2012). *Chasing Venus: The Race to Measure the Heavens.* Knopf, New York.

Young, T. (1845). *A Course of Lectures on Natural Philosophy and the Mechanical Arts.* Taylor and Watson, London.

Zeilinger, A. (2010). *Dance of the Photons.* Farrar, Straus and Giroux, New York.

Zoltán V. and Weihs G. (2015). "Foucault's method for measuring the speed of light with modern apparatus." *Eur. J. Phys.* **36**, 035013.

Zuo, J.M., Kim, M., O'Keeffe, M., and Spence, J.C.H. (1999). "Direct observation of d-orbital holes and Cu-Cu bonding in Cu_2O." *Nature* **401**, 49.

Zurek, W.H. (2002). "Decoherence and the transition from quantum to classical revisited." in *Los Alamos Science*, No. 27, p. 1.

Index

Figures are indicated by an italic *f* following the page number.

Index 241

Laue, Max von, 120–1
Le Gentil, Guillaume, 41–2
Le Verrier, Urbain, 88–9
Leggett, Tony, 207–8, 211
Leibnitz, Gottfried Wilhelm, 23–5
Lenard, Philipp, 173–4
length contraction, 134–8, 141–2, 150,
153–5, 228
length standard, 74–5, 180–1, 183–4
lenses, 7–8
Library of Alexandria, 32
lighthouses, 65–6
limb darkening effect, 45
linear superposition principle, 60
locality, 204–8, 210–12
Lodge, Oliver, 104–5, 134–8, 174–6,
178–9
longitude determination, 4, 18–21, 25–6,
40–1
Longitude Prize, 41, 44
longitudinal waves, 66–7, 92
Lorentz, Hendrik, 121–4, 129, 131–2, 134–5,
137–43, 154–5
Lorentz force, 142–3
Lorentz invariance, 154–6
Lorentzian transformations, 134–5, 140–1,
150, 154–5, 228
Lorenz, L.V., 175
lunar eclipses, 29, 30f, 31, 35
lunar parallax triangle, 31–2
Lyell, Charles, 189

magnetic fields, 92, 94–8, 95f, 136–7, 155,
160–1, 164
see also electromagnetism
magneto-optical effect, 92, 103
Mandelshtam, Leonid, 174
Manhattan atomic bomb project, 196,
209–10
many worlds interpretation of quantum
mechanics, 199–200
many-body problem, 100
Maraldi, Giacomo F., 21, 23–4
Marconi, Guglielmo, 18–19
Mars, 47, 191
Maskelyne, Neville, 41, 43–4
Mason, Charles, 41
mass standard, 180–1, 185

mathematical metaphors, 216–18
mathematical models, 217–19
matrix mechanics, 196
Maudlin, Tim, 208, 216
Maxwell, James Clerk, 14–15, 84–5, 91–2,
94–111, 102f, 104f, 108f, 114–17, 137–8,
159–60, 163, 181–2, 217
Maxwell, Katherine, 101, 102f
Maxwell's equations, 91–6, 101–12, 104f,
108f, 114, 119, 135–8, 140–1, 144–5, 150,
154–5, 164–6, 171–2, 175–6
measurement problem of quantum
mechanics, 199, 213–16, 214f
Mendel, Gregor, 191
Mercury
size and distance from Sun, 47
transit of, 38–40
meridian arc measurement, 74–5
Mermin, David, 206–7
metamaterials, 17
meter, definition of, 74–5, 180–1, 183–4
metric system of units, 180–5
Michelson, Albert, 57, 65, 84–5, 88, 112–17,
119–29, 126f, 127f, 131–2, 134–5, 137,
140–2, 144–6, 157, 200–1
microphones, 177–9
microwave ovens, 170–1, 229–30
microwaves, 170–2, 184
mirrors
specular reflection, 13–14, 13f, 81–2
see also rotating mirrors methods
mobile phones, 162, 184
molecular drugs, 197
Molyneux, Sam, 49–52, 51f
momentum, 156–7
Moon
eclipses, 29, 30f, 31, 35
size and distance to, 28–32, 29f, 30f, 34–5
moons of Jupiter, 4, 18–26, 22f, 116–17
Morley, Edward, 57, 112, 120–1, 123–4,
134–5
mutations, 191

natural selection, 190–1
Neptune, 47
Neumann, Carl, 165–6, 173
Neumann, John von, 209, 218
neural network algorithms, 217–18